OXFORD TEACHER REFERENCE

SECOND EDITION

DAILY WORKOUTS FOUNDATION TO YEAR 6

BUILDING MENTAL COMPUTATION SKILLS

ED LEWIS

NATASHA GILLARD

OXFORD

OXFORD
UNIVERSITY PRESS

Oxford University Press is a department of the University of Oxford. It furthers the University's objective of excellence in research, scholarship, and education by publishing worldwide. Oxford is a registered trademark of Oxford University Press in the UK and in certain other countries.

Published in Australia by
Oxford University Press
Level 8, 737 Bourke Street, Docklands, Victoria 3008, Australia.

First published 2004
Second edition 2021
Reprinted 2023

ISBN 978 0 19 033005 7

Editing consultancy by Clare Way
Illustrated by Marty Schneider, Martian Studio
Produced by Newgen KnowledgeWorks Pvt. Ltd.
Printed in China by Leo Paper Products Ltd

FSC
www.fsc.org
MIX
Paper | Supporting responsible forestry
FSC® C020056

Contents

Introduction

About Daily Workouts

Daily Workouts is a primary mathematics resource offering a comprehensive range of daily activities fostering mental computation and the development of mathematical skills, thinking and understandings. It is aligned with the Australian Curriculum: Mathematics, Number and Algebra strand and incorporates the proficiencies of understanding, fluency, problem-solving and reasoning.

What is mental computation?

Mental computation is a cognitive process used to produce an exact answer by mental methods, without the need for external devices such as pencil and paper or the electronic calculator – although pencils, paper and calculators can be and are used in this book to help develop mental skills. Mental computation requires the use of different strategies and instant recall of basic facts from memory in solving problems. The effect of manipulating numbers 'in the head' promotes the development of number sense, or a 'feel' for how numbers work.

Why use mental computation?

Mental mathematics is crucial for everyday living. It has been established that 80% of calculations made by adults are made mentally. Often estimation is needed to determine whether an answer obtained electronically is reasonable or not, for example, during shopping. *Daily Workouts* aids in the development of these necessary estimation skills.

How do I use *Daily Workouts* to teach mental computation?

Each activity in *Daily Workouts* has been designed as a teacher-led discussion to promote verbal communication among students, enhancing their thinking and reasoning. Activities have been designed primarily as lesson warm-ups, in which the teacher and students discuss and develop mathematical ideas. The activities may be also used flexibly during lesson breaks or for revision of previous concepts.

The activities in *Daily Workouts* are laid out in inquiry style with scripts of possible teacher questions and student responses/answers provided so that teachers can extend discussion and develop different ideas about the problem. Instead of drilling a fixed mental strategy in isolation, the teacher needs to explore a range of strategies suggested by the students, according to the type of activity given. In each case, the students are given a short time to investigate the problem concerned, before considering a variety of solution methods in classroom discussion with the teacher. The strategies that are most efficient are thus investigated and highlighted.

How do teachers assess students' mental computation strategies?

By allowing students enough time to discuss their mathematical thinking and not intervening too soon, the teacher can observe and note their responses, which provide a window to their thoughts. Requesting students to justify their strategies by asking questions such as 'Why?', 'How do you know that?' or 'Can you prove that?' can elicit further responses.

Skills in mental computation are thus best developed by students constructing their own knowledge from physical and mental activities in problem-solving contexts. The classroom emphasis should be on students' investigation, discussion and implementation of various mental methods, as their strategy of first choice when solving number problems, but answers are also provided for some problems.

How to use this book

Daily Workouts is designed to offer teachers a curated collection of activities to support mental computation and mathematical thinking. The activities are intended as 5- to 10-minute introductions or warm-ups in which the teacher and students are engaged in discussion and explanation.

Activities

Activities are organised by primary year group, strand, concept, topic and year level. A handy thumb tab helps to identify each year group – Lower, Middle and Upper. Supplementary activities can also be found on Oxford Owl, along with digital resources, including curriculum links, to support all of the activities.

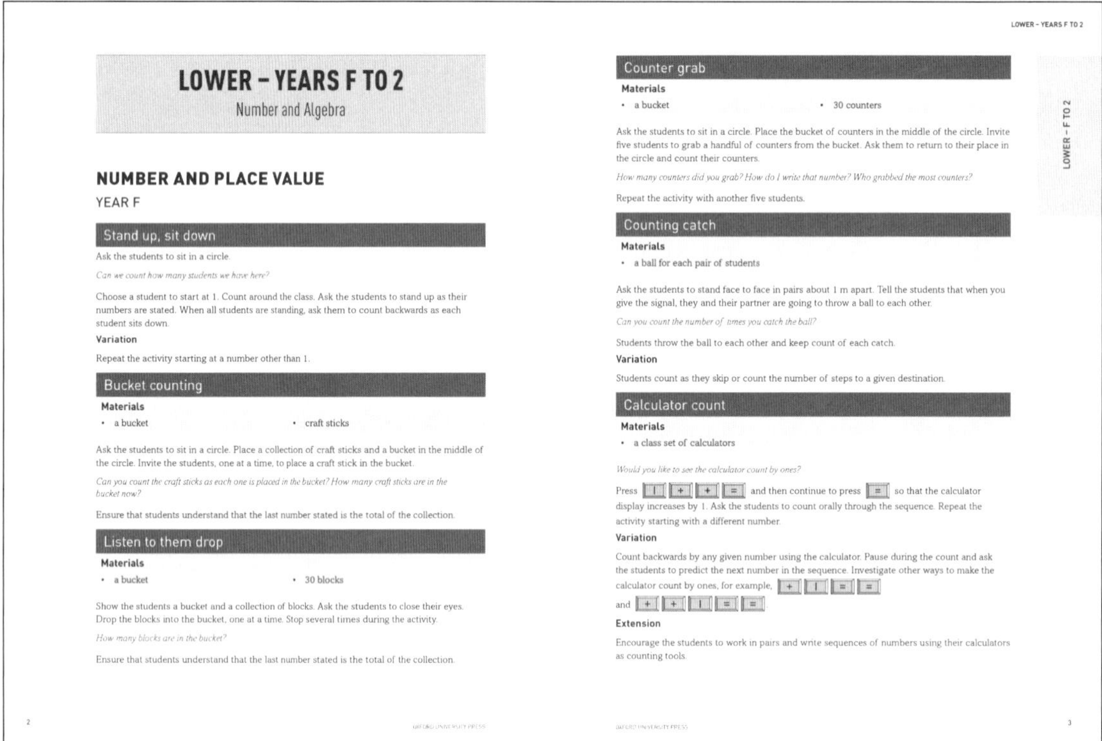

LOWER – YEARS F TO 2

LOWER – YEARS F TO 2

Number and Algebra

NUMBER AND PLACE VALUE

YEAR F

Stand up, sit down

Ask the students to sit in a circle.

Can we count how many students we have here?

Choose a student to start at 1. Count around the class. Ask the students to stand up as their numbers are stated. When all students are standing, ask them to count backwards as each student sits down.

Variation

Repeat the activity starting at a number other than 1.

Bucket counting

Materials

- a bucket
- craft sticks

Ask the students to sit in a circle. Place a collection of craft sticks and a bucket in the middle of the circle. Invite the students, one at a time, to place a craft stick in the bucket.

Can you count the craft sticks as each one is placed in the bucket? How many craft sticks are in the bucket now?

Ensure that students understand that the last number stated is the total of the collection.

Listen to them drop

Materials

- a bucket
- 30 blocks

Show the students a bucket and a collection of blocks. Ask the students to close their eyes. Drop the blocks into the bucket, one at a time. Stop several times during the activity.

How many blocks are in the bucket?

Ensure that students understand that the last number stated is the total of the collection.

2 OXFORD UNIVERSITY PRESS

Counter grab

Materials

- a bucket
- 30 counters

Ask the students to sit in a circle. Place the bucket of counters in the middle of the circle. Invite five students to grab a handful of counters from the bucket. Ask them to return to their place in the circle and count their counters.

How many counters did you grab? How do I write that number? Who grabbed the most counters?

Repeat the activity with another five students.

Counting catch

Materials

- a ball for each pair of students

Ask the students to stand face to face in pairs about 1 m apart. Tell the students that when you give the signal, they and their partner are going to throw a ball to each other.

Can you count the number of times you catch the ball?

Students throw the ball to each other and keep count of each catch.

Variation

Students count as they skip or count the number of steps to a given destination.

Calculator count

Materials

- a class set of calculators

Would you like to see the calculator count by ones?

Press [1] [+] [+] [=] and then continue to press [=] so that the calculator display increases by 1. Ask the students to count orally through the sequence. Repeat the activity starting with a different number.

Variation

Count backwards by any given number using the calculator. Pause during the count and ask the students to predict the next number in the sequence. Investigate other ways to make the calculator count by ones, for example, [+] [1] [=] [=] and [+] [+] [1] [=] [=].

Extension

Encourage the students to work in pairs and write sequences of numbers using their calculators as counting tools.

OXFORD UNIVERSITY PRESS 3

Materials

Each activity includes a list of materials to allow teachers to prepare in advance – it is assumed that everyday classroom materials such as paper, pencils, counters, Unifix or Multilink cubes and Base 10 materials are readily available.

Instructions

Each activity includes clear and simple instructions and a script featuring teacher dialogue and student responses that may lead to more class discussion and consideration of the problem or question. (Sometimes the answer is provided for quick reference, too.) Suggestions for variations and extension are also included where relevant.

How many counters?

Title

Materials

List of materials needed

- ten-frames

Briefly show a ten-frame with 7 dots:

Instructions

How many dots did you see? How do you know that there are 7 dots? (I saw 5 dots and I quickly counted 6, 7. I saw 5 dots and 2 dots and I worked it out on my fingers.)

Teacher dialogue and possible student responses

Diagram

Variation

Repeat the activity using different numbers represented on the ten-frames.

Variation activity

BLMs

All BLMs are available online at www.oxfordowl.com.au.

They include number cards, grids and templates, such as 3 × 3 grids, 4 × 4 grids, multiplication chart, hundreds chart, ten-frames, ten-strips and arrays.

DAILY WORKOUTS

Filled ten frames: 1–40 BLM 13

25 26

27 28

29 30

31 32

4 *Daily Workouts* © Oxford University Press 2021. This sheet may be reproduced for non-commercial classroom use.

LOWER – YEARS F TO 2

Number and Algebra

NUMBER AND PLACE VALUE

Year F

Stand up, sit down

Ask the students to sit in a circle.

Can we count how many students we have here?

Choose a student to start at 1. Count around the class. Ask the students to stand up as their numbers are stated. When all students are standing, ask them to count backwards as each student sits down.

Variation

Repeat the activity starting at a number other than 1.

Bucket counting

Materials

- a bucket
- craft sticks

Ask the students to sit in a circle. Place a collection of craft sticks and a bucket in the middle of the circle. Invite the students, one at a time, to place a craft stick in the bucket.

Can you count the craft sticks as each one is placed in the bucket? How many craft sticks are in the bucket now?

Ensure that students understand that the last number stated is the total of the collection.

Listen to them drop

Materials

- a bucket
- 30 blocks

Show the students a bucket and a collection of blocks. Ask the students to close their eyes. Drop the blocks into the bucket, one at a time. Stop several times during the activity.

How many blocks are in the bucket?

Ensure that students understand that the last number stated is the total of the collection.

Counter grab

Materials

- a bucket
- 30 counters

Ask the students to sit in a circle. Place the bucket of counters in the middle of the circle. Invite five students to grab a handful of counters from the bucket. Ask them to return to their place in the circle and count their counters.

How many counters did you grab? How do I write that number? Who grabbed the most counters?

Repeat the activity with another five students.

Counting catch

Materials

- a ball for each pair of students

Ask the students to stand face to face in pairs about 1 m apart. Tell the students that when you give the signal, they and their partner are going to throw a ball to each other.

Can you count the number of times you catch the ball?

Students throw the ball to each other and keep count of each catch.

Variation

Students count as they skip or count the number of steps to a given destination.

Calculator count

Materials

- a class set of calculators

Would you like to see the calculator count by ones?

Press [1] [+] [+] [=] and then continue to press [=] so that the calculator display increases by 1. Ask the students to count orally through the sequence. Repeat the activity starting with a different number.

Variation

Count backwards by any given number using the calculator. Pause during the count and ask the students to predict the next number in the sequence. Investigate other ways to make the calculator count by ones, for example, [+] [1] [=] [=] and [+] [+] [1] [=] [=].

Extension

Encourage the students to work in pairs and write sequences of numbers using their calculators as counting tools.

Counting circle

Ask the students to stand in a circle. Show these numbers: **10 20 30**

What do you notice about these numbers? (They all end in 0.)

Students repeatedly count around the circle from 1 to 30. Each time a student says '10', '20' or '30', that student sits down. Continue to count around the circle until only one student is left standing.

Take a counter

Materials

- 20 counters

Choose six students to sit in a circle. Place the counters in the middle of the circle and ask the students to count around the group, starting at 1. Each student says a number. If the student states a number with a '2' in it (2, 12, 20, 21, etc.), they collect a counter. When all the counters have been collected, the students count their own total.

How many counters do you have?

Repeat the activity with another group of six students.

Memory game to 10

Materials

- two sets of number cards from 1 to 10

Ask the students to sit in a circle. Place two sets of number cards from 1 to 10 face down in the middle. Choose two students to each turn over a card at the same time.

What is your number?

If the numbers are the same, the two students keep their cards. If not, they return them to the floor face down. Choose another pair of students and repeat the activity until all the cards are collected.

Number match

Materials

- a set of number cards from 1 to10 for each student
- a set of picture cards from 1 to 10
- 10 counters

Give each student a set of number cards from 1 to 10. Then hold up one card at a time in random order, using dots or pictures, to show 1 to10.

Can you show me the number card that is the same as this number?

Observe the students who need to count the dots one at a time and those that instantly recognise an arrangement for a particular number.

Can you show me the number card that is one more than this number?

Repeat the activity for all numbers from 1 to 10.

Count from. . .

Materials

- a bucket
- 13 Multilink or Unifix cubes

Show the students a bucket with 3 cubes in it. Tell the students that there are 3 cubes in the bucket.

Can you tell me how many cubes are in the bucket as I put more cubes in?

Continue to place 10 more cubes in the bucket, one cube at a time. When the students count to 13, remove the cubes from the bucket one at time and ask the students to count backwards by ones. Repeat the activity often. Start with a different number of cubes each time.

Counting fingers

Ask the students to work with a partner for this activity.

Can you and your partner show me 18 fingers altogether?

Allow the students some time to represent 18 fingers.

Can you show me 15 fingers? Can you show me 15 fingers another way?

Encourage the students to count to check the total number of fingers.

Clap and chant

Create a slow rhythmic pattern by clapping your hands then tapping your knees. Encourage students to continue the clapping and tapping pattern. On a 'clap', say a number from 1 to 20, such as 16.

Say the number that comes after the number 16 on the next 'clap'. (17)

Repeat the activity. Choose a number and ask the students to say the number after it on the next clap.

Variation

Repeat the activity for numbers that come before the chosen number.

Compare with five

Materials

- a five-strip
- counters

Show a blank five-strip. Invite a student to place 5 counters on the five-strip.

How many counters can you see?

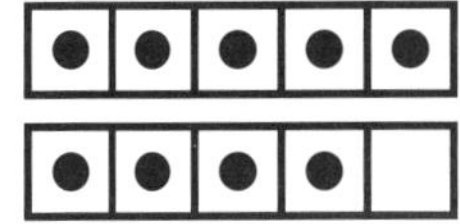

Show another blank five-strip. Invite a student to put 4 counters on this strip. Show the counters for 3 seconds:

What did you see? How many counters are there altogether? How do you know there are 9 counters? (I saw 5 counters and I quickly counted 6, 7, 8, 9. I saw 5 and 4 counters.)

Repeat the activity for numbers between 5 and 10.

Making teens

Show these numbers: **15 18 19 16**

What do you notice about these numbers? (They all start with the number 1. When you say them you say 'teen'.)

Choose two students to stand in front of the class. Ask the first student to show 10 fingers.

How many fingers does (name of the second student) need to hold up to show 15 fingers altogether?

Ask the two students to show 15 fingers. With the class, count all 15 fingers to check the total. Ask the students to complete this sentence: '15 is 10 and (5) more'. Choose other pairs of students to show 16, 18 and 19 fingers.

Memory game to 20

Materials

- two sets of number cards from 11 to 20

Ask the students to sit in a circle. Place two sets of number cards from 11 to 20 face down in the middle. Choose two students to each turn over a card at the same time.

What is your number?

If the numbers are the same, the two students keep their cards. If not, they return them to the floor, face down. Choose another pair of students and repeat the activity until all the cards are collected.

Dice patterns

Materials

- dot dice cards or a dot dice

Briefly show one of the cards, such as the one for 5:

How many dots did you see? What did they look like? (I saw 4 and 1 more in the middle.) How many dots will be there if 1 more was added?

Encourage students to state the number of dots instantly (without counting each dot one by one). Discuss the students' responses. To check, show the arrangement again. Repeat the activity often using other dot dice cards.

Number strips

Materials

- counters

Show a number strip with counters placed as follows:

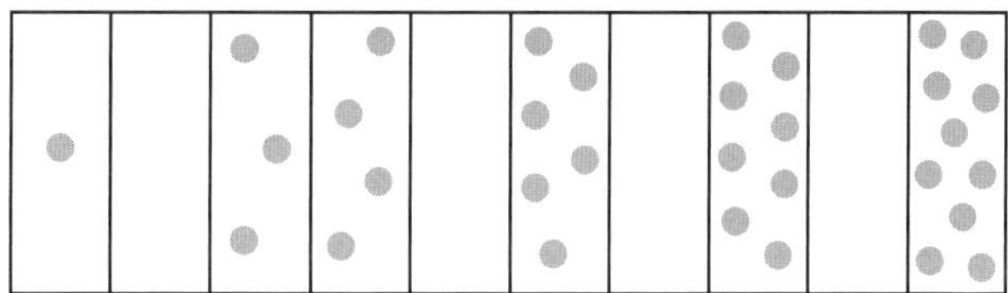

Which numbers are missing?

Discuss the students' responses. Show or add the missing counters to complete the pattern. Ask the students to work in pairs to construct the pattern with counters.

Variation

A set of random dot cards could also be used for this activity. Students can supply the missing cards in the sequence.

More dots

Materials

- a set of random dot cards from 1 to 10

Show and discuss the terms *more* and *less*. Shuffle the dot cards and place them face down in a pile. Turn over the top two cards and show them to the students.

Which card has more dots? How can we check? (We can count the dots.)

Ask the students to count the number of dots on each card. Encourage students to make statements using the terms *more* or *less*.

How many more than 3 does 7 have? How did you work it out? (7 is 4 more than 3 and 3 is 4 less than 7)

Discuss the students' responses. Repeat the activity often using other dot cards.

Estimating dots

Materials

- a set of random dot cards from 1 to 10

Ask the students to close their eyes and imagine 6 dots.

Can you describe what you see? (I imagined two lines of 3 dots, like a dice.)

Show the random dot cards one at a time:

Which one of these cards has 6 dots on it?

Tell the students to raise their hands when they see a card with 6 dots. Repeat the activity for cards showing other numbers of dots.

Variation

Repeat the activity with cards showing more than 10 dots.

More or less?

Materials

- two sets of number cards from 1 to 10

Show this number line:

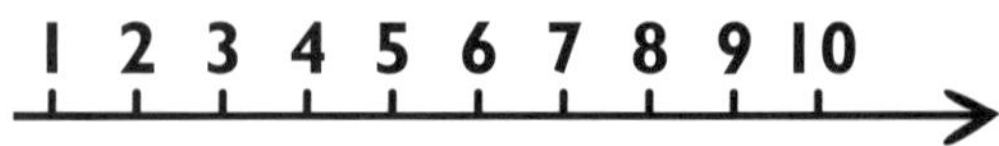

Which numbers are more than 7? Which numbers are less than 4? How many numbers are less than 8?

Shuffle both sets of the number cards and place them face down in two piles (a teacher pile and a student pile). Turn over the top card on the teacher pile and locate the number on the number line. Invite a student to predict whether the top card on their pile will be more or less than the teacher's card. The student then turns over the next card to check their prediction. Continue the activity until all cards are used.

Ordering numbers

Materials

- a set of number cards from 0 to 10

Give 11 students each a number card from 0 to 10. Ask the students with the cards for 0 and 10 to stand at opposite ends of the room. One at a time, invite the students with the remaining cards to stand where their number is approximately located between 0 and 10.

How do you know where to stand? Which numbers will stand next to you? (Number 2 is closer to 0.)

Continue until all students with number cards have located themselves between 0 and 10.

Before and after

Material

- a dot dice

Ask the students to sit in a circle. Roll a dot dice into the middle several times and ask the students to state the number rolled each time.

What is the number that is 1 more than the number rolled? What is the number that is 1 less than the number rolled? How do you know?

Discuss the students' responses.

Before and after cards

Materials

- a set of number cards from 1 to 10 for each student

Give each student a set of number cards from 1 to 10.

Can you show me the number that is one less than 9? Can you show me the number that is one more than 3?

Count from 1 to check students' answers. Repeat this process for other numbers.

Can you show me a number that is more than 4? Can you show me a number that is less than 4?

Discuss all the possible answers in the range from 1 to 10. Repeat this process for other numbers.

Variation

Repeat the activity for numbers 11 to 20.

Comparing towers

Materials

- 12 yellow Multilink or Unifix cubes
- 7 blue Multilink or Unifix cubes

Show the students a tower of 12 yellow cubes and a tower of 7 blue cubes:

Which tower has more cubes? How can we check?

Discuss the towers in terms of their height and the number of cubes used to construct them.

How many more cubes have been used to make the taller tower? (There are 5 more cubes on the yellow tower.)

Discuss the students' strategies.

Variation

Repeat the activity using towers with different numbers of cubes.

Imagining numbers

Show this number line:

Tell the students that the line starts at 0 and goes up by ones.

Where does number 7 go? Where is number 12?

Record number 7 and number 12 on the line.

Where does number 11 go? Is it closer to 7 or 12? What number comes after 2? . . . 7? . . . 12?

Record these numbers on the number line.

Which numbers are missing from the number line?

Ask the students to describe where these numbers sit in relation to the existing numbers. *(Number 4 goes after number 3 . . .)*

Complete the labelling of the number line.

In the range

Show an open number line from 0 to 15:

Can you tell me a number that is in between these numbers? Is that number closer to 0 or 15?

Mark the number on the number line. Continue this process until all whole numbers have been placed on the line.

Variation

Repeat this activity for other number lines.

How many counters?

Materials

- ten-frames

Briefly show a ten-frame with 7 dots:

How many dots did you see? How do you know that there are 7 dots? (I saw 5 dots and I quickly counted 6, 7. I saw 5 dots and 2 dots and I worked it out on my fingers.)

Variation

Repeat the activity using different numbers represented on the ten-frames.

Teen towers

Materials

- 10 Multilink or Unifix cubes of one colour
- 10 Multilink or Unifix cubes of another colour

Make a tower of ten cubes:

How many cubes are there? How many more cubes do I need to make 17 cubes altogether?

Add 7 more cubes to the tower, one at a time. Ask the students to count as you join each cube.

17 is made of 10 and . . .? How many cubes do I need to take away to leave 10 cubes?

Remove 7 cubes from the tower, one at a time. Ask the students to count as you remove each cube.

Variation

Repeat the activity for other teen numbers.

Numbers ahead

Show a vertical number line from 0 to 20:

20 19 18 17 16 15 14 13 12 11 10 9 8 7 6 5 4 3 2 1 0

Choose a student to stand in front of the room facing the class. Write a number from 0 to 20, above the student's head, such as 19. The student attempts to guess the number. The class responds by saying if the guessed number is *lower* or *higher* than the number above the student's head.

Missing number

Materials

- a set of number cards from 10 to 20

Ask the students to sit in a circle. Place a set of number cards from 10 to 20 face up in the middle. Invite two students to stand up.

Can you place these cards in order from 10 to 20?

After the students have placed the cards in order, ask them to close their eyes. Remove one number card and rearrange the remaining cards. Ask the students to look at the cards to determine which number is missing.

Can you tell me which number is missing? How did you work it out?

Discuss the students' strategies. Repeat the activity with another pair of students.

Number card line

Materials

- a set of number cards from 1 to 20

Invite some students to place the number cards on the floor in order from 1 to 20. Turn the number cards face down, maintaining their order.

Where is number 15? How did you work it out?

Discuss the location of number 15. Check by counting forwards from 1 and backwards from 20. Turn the card over to reveal number 15.

What is the number that is 2 more than 15? How did you work it out?

Discuss the students' strategies.

Variation

Repeat the activity for other numbers from 1 to 20.

Mixed numbers

Materials

- 5 blank cards

Choose five students to tell you a number.

Can you tell me a number that you say when you count to 20?

Record each number on a separate card. Give each student their number to hold and show to the class.

Can you stand in order from the lowest to the highest number?

Ask the class to count by ones from 0 to check the order. Repeat the activity with different students.

Larger range

Show an open number line with 12 at one end and 25 at the other:

Ask a student to state a number.

12 ——————————————————— **25** →

Can you tell me a number that is in between these numbers? Is that number closer to 12 or 25?

Write the number on the number line. Continue this process until all whole numbers have been placed on the line.

Variation

Repeat this activity for other number lines.

Estimating groups

Materials

- 40 counters

Place a collection of approximately 10 counters in a group.

How many counters do you think are in this group?

As a class, count the counters in the group. Then place approximately 30 counters in a second group.

How many counters do you think are in this group? Is there more or less than 10 counters?

Call for estimates and then count the group.

Which group has more counters? How do you know? (30 is more than 10. When you count, you say 10 before you say 30.)

Discuss the students' responses.

Whose number?

Materials

- a set of number cards from 1 to 30

Ask the class to sit in a circle. Give each student a number card from 1 to 30. Tell the students that you are going to ask them some questions about their number cards. Students whose numbers fit the questions can stand up.

Who has a number that comes after 16? Who has a number that comes before 9? Who has a number with only 1 digit? Whose number has 2 digits? Whose number has 2 digits that are the same?

Discuss the students' responses. Refer students to a hundreds chart or number line if required.

Where is the number?

Show a number line from 1 to 30, with intervals of 10:

How can you work out where number 23 would be? Is 23 closer to 20 or 30? How do you know? (It is closer to 20 because it is only 3 away. 23 is between 2 tens and 3 tens.)

1 10 20 30

Discuss the students' responses. Mark the position of 23 on the number line. Repeat the activity for other numbers on the number line.

Between decades

Show a number line from 1 to 30, with intervals of 10:

Point to a mark on the number line.

1 10 20 30

Which number am I pointing to? How did you work it out? (It is 6 more than 20 so it is 26. It is 4 less than 30 . . . 29, 28, 27, 26.)

Discuss the students' responses. Write the number on the number line. Repeat the activity for other numbers on the number line.

Extension

Repeat the activity for other decades when students are ready.

Count to 30

Materials

- ten-frames
- pencils
- paper

Show these ten-frames:

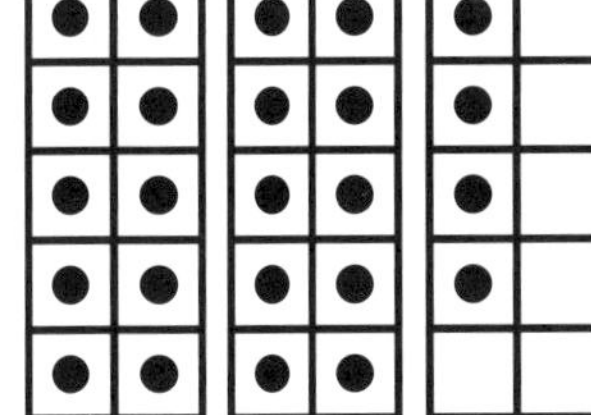

How many dots are there?

Ask the students to record the number of dots.

How do you know that there are 24 dots? (I counted them one at a time. I saw 20 and I counted 4 more.)

Discuss the students' responses.

Variation

Repeat the activity using other numbers to 30.

Hidden number

Materials

- a number grid
- a counter

Cover the number 24 on a number grid:

Which number have I covered? How do you know? (24 comes after 23. It is 10 more than 14.)

1	2	3	4	5	6	7	8	9	10
11	12	13	14	15	16	17	18	19	20
21	22	23	●	25	26	27	28	29	30
31	32	33	34	35	36	37	38	39	40
41	42	43	44	45	46	47	48	49	50

Encourage students to give as many reasons as they can for determining the covered number. Invite students to place counters on the chart and repeat the activity.

Numbers around

Materials

- a number grid
- a counter

Cover the number 37 on a number grid.

Which number have I covered? How do you know? Which number is one more than the number I have covered? Which number is one less than the number I have covered?

Discuss the students' responses. Repeat the activity for other numbers from 1 to 50.

Extension

Repeat the activity to identify numbers that are 10 before and 10 after a given number.

Forwards count

Materials

- a number grid
- a counter

Cover the number 12 on a number grid (see example on page 13).

Which number will I land on if I move the counter forwards by 6?

Ask the students to count as you move the counter to number 18.

Which number will I land on if I move the counter forwards by 4?

Invite a student to move the counter, one number at a time, to check the answer. Repeat the activity starting at other numbers on the chart.

Backwards count

Materials

- a number grid
- a counter

Cover the number 12 on a number grid (see example on page 13).

Which number will I land on if I move the counter backwards by 2?

Ask the students to count as you move the counter to number 8.

Which number will I land on if I move the counter backwards by 4?

Invite a student to move the counter, one number at a time, to check the answer. Repeat the activity starting at other numbers on the chart.

Subitise and add

Materials

- dot dice cards from 1 to 6

Briefly show a dot card to the students:

Can you tell me how many dots you see?

Encourage the students to state the number of dots instantly (without counting each dot one by one). When the students are proficient at this, show them two dot cards at once:

How many dots are there altogether? How did you work it out? Is there a quicker way of working out the total number of dots? (I counted on from 5. That is 6, 7. I put up 5 fingers and 2 more fingers to make 7 fingers.)

Discuss and share the students' strategies.

Subitise and subtract

Materials

- dot dice cards from 1 to 6

Show the dot card for 6:

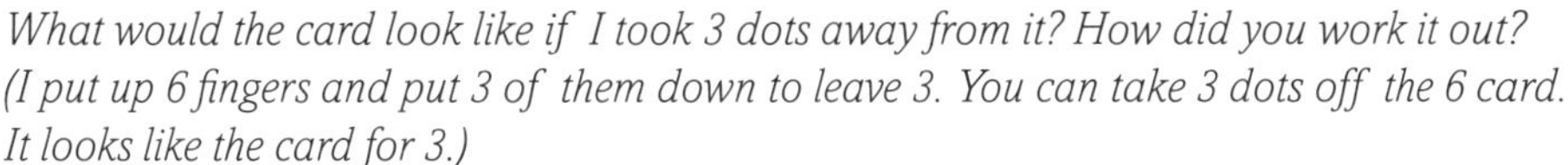

What would the card look like if I took 3 dots away from it? How did you work it out? (I put up 6 fingers and put 3 of them down to leave 3. You can take 3 dots off the 6 card. It looks like the card for 3.)

Discuss and share the students' strategies.

What is the difference?

Materials

- a dice
- 50 Multilink or Unifix cubes

Ask the students to sit in two teams. Ask a student from the first team to roll the dice and state the number rolled.

Can you make a tower using that number of cubes?

Choose a student from the second team to repeat the activity. Compare the two towers to determine the difference.

Whose tower has the most cubes? How many more cubes does this team have?

The team with the larger number keeps the number of cubes that is the difference.

Repeat the activity until one team collects a total of 15 cubes.

Different dots

Materials

- ten-frames

Show 2 ten-frames with 10 dots in one and 6 in the other:

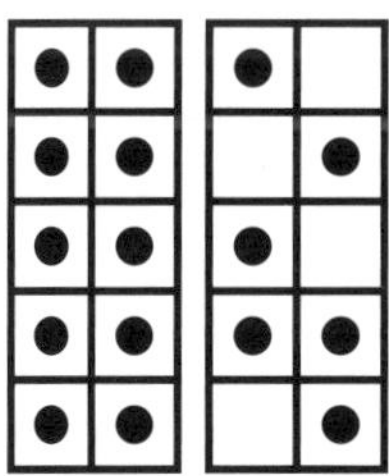

How many dots can you see in each frame? Which frame has more dots? How many more?

Discuss the students' strategies. Focus on the idea of comparison in the discussion. Repeat the activity for other numbers from 0 to 10.

Year 1

Counting bucket

Materials

- a bucket
- craft sticks

Tell the students that there are 47 craft sticks in the bucket.

Can you count on as I put some more sticks in the bucket?

Continue to place 12 more craft sticks in the bucket, one at a time. Ask the students to keep count.

(47, 48, 49, 50, 51 . . .)

When the students count to 62, remove the sticks from the bucket one at a time as students count backwards by ones. Repeat the activity often. Start with a different 2-digit number each time.

Craft stick tens

Materials

- a bucket
- 9 bundles of 10 craft sticks

Place two bundles of 10 craft sticks in a bucket. Tell the students that there are 20 craft sticks in the bucket. Continue to place bundles of 10 sticks in the bucket, asking students to count by tens.

Can you help me to count the number of sticks in the bucket? (30, 40, 50, 60 . . .)

Count by tens to 90. At 90, remove bundles of 10 craft sticks from the bucket, one by one. Ask the students to continue counting backwards by tens. Repeat the activity, starting with different numbers of bundles of 10 craft sticks.

Count from 15

Materials

- a bucket
- Multilink or Unifix cubes

Show the students a bucket with 15 cubes in it. Tell the students that there are 15 cubes in the bucket.

Can you tell me how many cubes are in the bucket as I put more cubes in?

Place two cubes in the bucket each time. Encourage students to count as you add each pair of cubes.

(15, 17, 19, 21, 23, 25 . . .)

After reaching 25, remove pairs of cubes from the bucket, one at a time. Ask the students to count backwards by twos.

(25, 23, 21, 19 . . .)

Repeat the activity often. Start with a different number of cubes each time.

Extension

Repeat the activity, counting forwards and backwards by fives or tens.

Twos count forwards and back

Materials

- a class set of calculators

Would you like to see the calculator count by twos?

Enter [2] [+] [=] and [=] and then continue to press [=] so that the calculator display increases by 2. Ask the class to count through the sequence. Pause during the count and ask individual students to predict the next number in the sequence. Repeat the activity using a different number to start.

Would you like to see the calculator count backwards by twos?

Enter [5] [0] [−] [2] and [=] and then continue to press [=] so that the calculator display decreases by 2. Ask the class to count orally through the sequence. Repeat the activity using a different starting number.

Extension

Practise counting forwards and back by fives or tens using the calculator as a counting tool. Encourage students to work in pairs and write number sequences on paper or cardboard strips.

Count the lollies

At a party I saw 4 cupcakes on a plate. Each cake had 2 lollies on top. How many lollies were on the plate altogether?

Encourage the students to close their eyes and visualise the groups.

Can you picture the cakes and the lollies in your mind? Can you tell me how many lollies you can see? (2 . . . 4, 5, 6, 7, 8. . . 2, 4, 6, 8 . . . altogether.)

Discuss the students' counting strategies. Show 4 groups of 2 lollies and model as equal groups.

Jumping Joey

Show a number line from 0 to 20: 0 ——————————— 20

Jumping Joey was playing on a number line. He started at 0 and kept jumping forwards by twos. What numbers will Jumping Joey land on? How many jumps will Joey make to get to number 20?

Invite a student to record the jumps on the number line.

What do you notice about the numbers that Joey lands on? (He is jumping on every second number. The numbers are even.)

Discuss the students' observations.

Counter points

Materials

- red and blue counters
- a bag

Show the students a bag of mixed red and blue counters. Invite a student to place their hand in the bag to remove a handful of counters.

How many blue counters do you have? How many red counters do you have? (I have 10 red counters and 8 blue counters.)

If red counters count for 1 point and blue counters count for 2 points, how many points is [the student] holding? How did you work it out? (I counted 2, 4, 6, 8 . . . 16, plus 10. I doubled the number of blue counters, then added 10 more points.)

Share the students' strategies.

Jump by fives

Show a number line from 0 to 50, with intervals of 5:

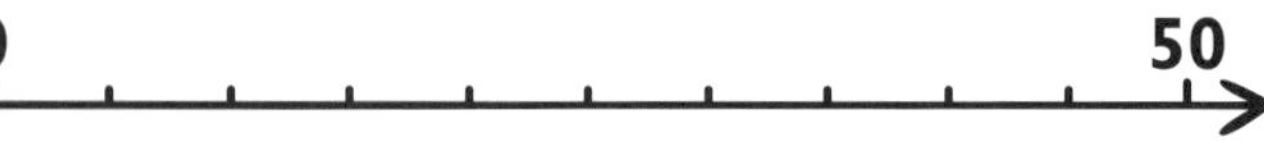

Invite the students to count by fives from 0 to 50.

Where does number 35 go? Where does number 15 go? If I started at 0, how many jumps of 5 would it take to get to 50? If I started at 20, how many jumps of 10 would it take to get to 50? If I was at 10 and jumped forward three more jumps of 5, where would I land? (15 . . . 20 . . . 25. That is 3 jumps of 5 from 10.)

Discuss the students' strategies for obtaining answers.

Handful of fives

Materials

- a bucket
- Multilink or Unifix cubes

Divide the class into three teams. Choose a student from each team to grab a handful of cubes from the bucket.

How many cubes did you grab? If each cube is worth 5 points, how many points did you earn? Who scored the most points?

The student with the greater number of points earns a point for their team. Repeat the activity until all students have had a turn.

Buzzing Bee

Show a number line from 0 to 30:

0 1 2 3 4 5 6 7 8 9 10 11 12 13 14 15 16 17 18 19 20 21 22 23 24 25 26 27 28 29 30

A buzzing bee was playing on a number line. It started at 30 and kept jumping backwards by fives. What numbers will the bee land on? How many jumps will the bee make to get to 0?

Invite a student to record the jumps on the number line.

Variation

Repeat the activity, starting the bee at different multiples of 5.

Decade counting

Ask the students to stand in a circle. Show these numbers: **10 20 30 40 50**

What do you notice about these numbers? (The first number goes 1, 2, 3, 4, 5. They are the numbers you say when you count by tens.)

Discuss the students' responses.

Can you count around the circle by tens?

Each time a student says '100', that student sits down. The count then begins again. Continue the activity until only one student is left standing.

Variation

Repeat the activity counting backwards by tens from 100. The student who says '0' (zero) sits down.

Count by tens

Materials

- 1 ball for each pair of students

Ask the students to stand face to face in pairs, about 1 m apart. Tell the students that when you give the signal they are going to throw a ball backwards and forwards to each other.

Can you count by tens as each person catches the ball?

As students throw the ball to each other, they count 10, 20, 30, 40 . . . . Repeat the activity, counting by twos or fives.

Variation

Students count by tens, fives or twos as they skip.

Estimate by 10

Materials

- 10 counters traced end to end on strips of card
- a whiteboard, wall or shelf

How many counters do you think will fit along the length of the board/wall/shelf? What is your best guess (I think 80 counters will fit.)

Show the students a strip of card showing the outline of 10 counters: 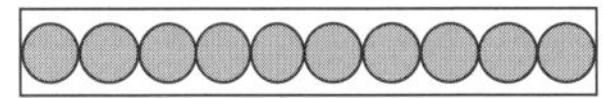

How many strips of 10 counters do you think will fit along the length of the board/shelf/wall? How many counters is that?

Place the strips of 10 counters end to end along the length of the board/shelf/wall. Allow students to change their estimation as each strip is placed. Count by tens to determine the actual length of the board/shelf/wall. Repeat the activity, measuring the lengths of different objects in the room.

Estimating groups

Materials

- 80 craft sticks

Show the students a pile of 80 craft sticks.

How many groups of 10 craft sticks can we make with these sticks? What is your best guess? (I think there will be 8 groups.)

Invite students one at a time to count and hold 10 craft sticks.

How many groups of 10 have we made? How many craft sticks is that?

Discuss the students' responses.

Popcorn problem

Materials

- 100 popcorn kernels and 10 cups

Show the students 100 popcorn kernels and 10 cups.

If we have 100 pieces of popcorn, how many cups of 10 pieces can we make?

Encourage students to first estimate the groups of 10. Choose one student at a time to count 10 popcorn kernels into a cup.

How many cups of 10 popcorn kernels did we make? How many kernels do I need to make 12 cups?

Discuss the students' strategies.

Off decade patterns

Materials

- a hundreds chart

Show the students a hundreds chart. Invite a student to place a mark on number 3.

What is the number that is 10 more than 3?

Count on 10 to 13. Place a mark on 13. Repeat several times.

What is the number that is 10 more than 13? 10 more than 23? 10 more than 33? . . .

Continue to place marks on the chart:

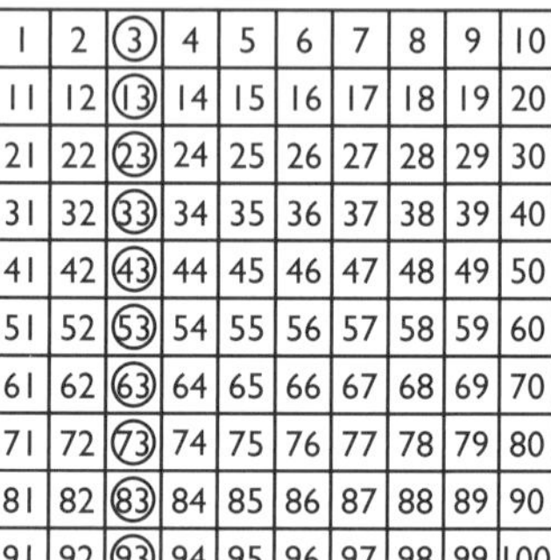

1	2	3	4	5	6	7	8	9	10
11	12	13	14	15	16	17	18	19	20
21	22	23	24	25	26	27	28	29	30
31	32	33	34	35	36	37	38	39	40
41	42	43	44	45	46	47	48	49	50
51	52	53	54	55	56	57	58	59	60
61	62	63	64	65	66	67	68	69	70
71	72	73	74	75	76	77	78	79	80
81	82	83	84	85	86	87	88	89	90
91	92	93	94	95	96	97	98	99	100

What patterns do you notice? (The first number goes 1, 2, 3, 4 . . . like that. The numbers all have a 3 in them.)

Discuss the students' responses. Remove the hundreds chart.

What is the number that is 10 more than 43? . . . 73? . . . 33? . . . 83? How do you know?

Buzz off decade

Ask the students to stand in a circle. Show these numbers: **13 23 33 43 53**

What do you notice about these numbers? (The first number goes 1, 2, 3, 4, 5. They all end in 3.)

Discuss the students' responses. Ask the students to count by tens around the circle starting at 3. *(3, 13, 23, 33, 43, 53, . . . 93)*

Each time a student says '93', that student sits down. The count then begins again. Continue the activity until only one student is left standing. Repeat the activity, starting the count from a different 1-digit number.

Whose number?

Materials

- a set of number cards from 1 to 30
- a hundreds chart

Ask the students to sit in a circle. Give each student a number card from 1 to 30. Ask students with the numbers that match these questions to stand and show their cards.

Who has a number that you say when you count by fives from 0? Who has a number that you say when you count by tens from 0? Who has a number that you say when you count by tens from 2?

Refer students to the hundreds chart to check the answers. Repeat the activity for numbers from 30 to 60.

Craft sticks groups

Materials

- 3 bundles of 10 craft sticks

Ask the students to sit in a circle. Place 3 bundles of 10 craft sticks in the middle of the circle.

If we put the craft sticks in groups of 5, how many groups will we have? What do we have to do first? (We need to break up the bundles.)

Choose students to group the craft sticks.

How many groups of 5 craft sticks do we have? How many craft sticks are there altogether?

Discuss the students' responses.

Craft stick race

Materials

- a bucket
- craft sticks (some bundled into tens)

Ask the students to sit in a circle. Place the bucket of craft sticks in the middle. Choose three students to play the game.

Can you show me 24 craft sticks?

Ask each student to race to the collection of craft sticks and quickly arrange 24 craft sticks on the floor.

How did you make number 24? Can you tell me how you count to check the number of craft sticks? (It is too slow to pull them out one by one. 10, 20, 21, 22, 23, 24 . . .)

Ask the first student to arrange 24 craft sticks to remain standing. Choose two new competitors. Repeat the activity, asking the students to represent a different number of craft sticks.

Comparing bundles

Materials

- 6 bundles of 10 craft sticks
- 11 single craft sticks

Show the students the first collection of two bundles of 10 craft sticks and 8 single craft sticks:

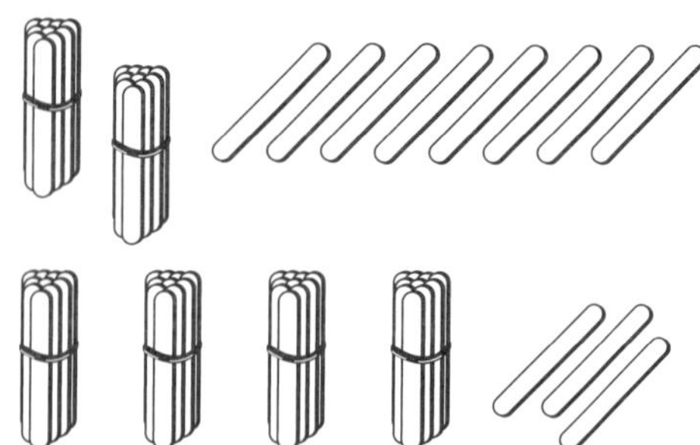

Show a second collection of four bundles of 10 craft sticks and 3 single craft sticks:

Which collection has more craft sticks? How do you know? (43 has 4 tens and 28 has 2 tens, so 43 has more. 43 is more than 28.)

Discuss the students' responses.

How can you check to make sure?

Repeat the activity using different craft stick collections.

Thirties tower

Materials

- 40 Multilink or Unifix cubes (10 cubes of 4 different colours)

Make a tower of 30 cubes, using 10 cubes of 3 different colours.

How many cubes are there? How many more cubes do I need to make 36 cubes altogether?

Add 6 more cubes to the tower, one at a time. Ask the students to count as you add each cube.

36 is made of 30 and . . .? How many cubes do I need to take away to leave 6 cubes?

Remove a colour group of 10 cubes at a time. Ask the students to count as you remove each colour group, leaving 6 cubes. Extend the activity to higher decades.

Fingers show

Ask the students to sit in groups of four. Show this number: **36**

Can your group show me 36 fingers? (3 people can make 30 fingers, and then there are 6 more. We have 10 and 10 and 10 and 6.)

Invite one of the groups to justify that they have 36 fingers.

How many more fingers are needed to make 40 fingers altogether? How did you work it out? (10 plus 10 plus 10 plus 10 is 40)

Repeat the activity. Ask the students to show a different number of fingers. Determine the number of fingers needed to make 40.

Compare towers

Materials

- 100 Multilink or Unifix cubes (in different colours)

Make a tower of 37 cubes and another of 42 cubes. Use contrasting colours for each group of 10 cubes in the towers.

Which tower has more cubes? How can we check? (You could put them next to each other.)

Discuss the towers in terms of their height (or length) and the number of cubes used to construct them.

How many more cubes have been used to make the taller tower? (You can see that there are 5 more cubes on the longer tower.)

Discuss the students' strategies. Repeat the activity using towers with different numbers of cubes.

Before and after

Materials

- a class set of calculators

Enter [3] and [8] on the calculator display.

What is the number that is 1 more than this number? What is the number that is 1 less than this number? How can we check the answers using the calculator?

Show the students how to enter [3] [8] [–] [1] [=] and [3] [8] [+] [1] [=] on the calculator to check. Repeat for other 2-digit numbers.

Where do you stand?

Materials

- a set of number cards from 0-50
- 6 beanbags

Mark out a number line from 0 to 50 in the playground, with intervals of 10. Ask each student to select a number card.

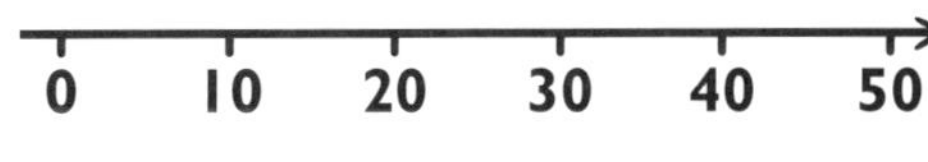

Who has a number between 20 and 30? Can you read your number to me? Can you stand on the number line where your number is located?

Repeat the activity for each decade of numbers (0 to 10, 10 to 20, 30 to 40, and 40 to 50).

Higher or lower?

Show a vertical number line from 0 to 50:

Choose a secret number in this range. Invite students to guess the number.

I am thinking of a number between 0 and 50. Can you guess my number? (Is it 30?)

My number is higher/lower than . . .

Mark each guess on the number line. Continue until the chosen number is determined. Repeat the activity using other number lines.

Between decades

Show this number line:

0 10 20 30 40 50 60 70 80

Tell the students that the line starts at 0 and goes up by tens.

Which number is halfway between 40 and 50? Where does number 23 go? Is it closer to 20 or 30? Which numbers are between 30 and 40? Is number 43 more than or less than 66? How do you know?

Share and discuss the students' answers.

Numbers between

Show this number line:

50 100

Can you tell me a number that is in between these numbers?

Discuss the students' suggestions. Choose a student and ask him or her to state a number, for example, 58.

Is the number closer to 50 or 100? How do you know? (58 is only 8 away from 50 so it is closer to 50 than 100.)

Mark the number on the number line. Repeat the activity. Ask the students to describe and justify the location of another ten numbers on the number line.

Imagining numbers

Show this number line:

Where does number 90 go? Where does number 110 go? Where does number 37 go? Is it closer to 20 or 40? (If you add on 3 to 37 you get 40. So it is very close to 40.) Which numbers are between 60 and 80?

Record the students' responses. Ask the students to explain their answers.

Ordering numbers

Materials

- string
- cardboard
- pencils
- pegs

Give each student a piece of cardboard and ask them to write a number between 0 and 100 on it.

Make a number line by tying a length of string across the room. Place 0 at one end and 100 at the other.

Invite students to place their number card in the appropriate position on the line.

Can you read your number to me? Is your number closer to 0 or 100? What is the number before/after your number?

Number line strips

Materials

- a cardboard strip
- a peg for each student

Give each student a cardboard strip and a peg. Ask him or her to record 0 at one end and 200 at the other.

Can you place the peg where number 50 would be on the number line? Can you describe where you have placed your peg? Why did you place the peg in that location?

Continue to ask the students to locate various numbers on the number strip. Encourage them to explain their placement of the peg.

Dots and spaces

Materials

- ten-frames
- Show these ten-frames:

How many dots can you see? How many spaces are there? How did you work it out? (I saw 10 and more 8. I saw 5 and 5 and 3.)

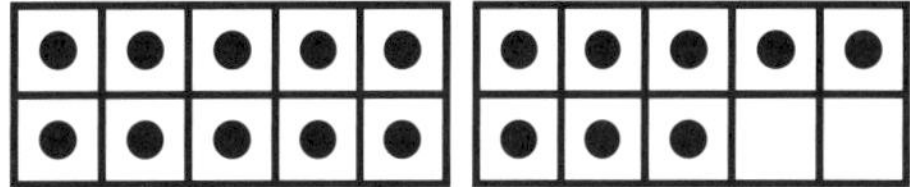

Discuss the students' responses. Check or count the number orally with the students. Repeat the activity for other numbers to 30.

Which number?

Materials

- a hundreds chart

Show the students a hundreds chart.
Use a counter to cover a number on the chart:

What is this number on the hundreds chart? How do you know? (83 is 10 more than 73. 83 comes after number 82.)

Discuss the students' responses. Repeat this activity often by changing the location of the counter on the hundreds chart.

1	2	3	4	5	6	7	8	9	10
11	12	13	14	15	16	17	18	19	20
21	22	23	24	25	26	27	28	29	30
31	32	33	34	35	36	37	38	39	40
41	42	43	44	45	46	47	48	49	50
51	52	53	54	55	56	57	58	59	60
61	62	63	64	65	66	67	68	69	70
71	72	73	74	75	76	77	78	79	80
81	82		84	85	86	87	88	89	90
91	92	93	94	95	96	97	98	99	100

Ten more or less

Materials

- a hundreds chart
- paper
- pencils

Show the students a hundreds chart.

Can you write the number that is 10 more than 49?

Check the answer by counting by ones on the hundreds chart from 49. Draw the students' attention to the location of the answer.

Look at 59. It is right underneath 49. Can you write and show me the number that is 10 more than 26?

Review the students' answers. Check by counting along the hundreds chart.

Variation

Repeat the activity for numbers that are 10 less than a given number.

Move it on

Materials

- a hundreds chart
- a counter

Show the students a hundreds chart. Put a counter on number 38.

What number will I land on if I move the counter forwards by 7? (38 add 2 is 40, add 5 more makes 45.)

Ask the students to count as you move the counter to number 45.

What number will I land on if I move the counter forwards by 6?

Invite a student to move the counter to determine the answer. Repeat the activity starting at other numbers.

Ten more

Materials

- number cards for 24, 13, 36, 25 and 58

Show the students one card at a time.

What is the number that is 10 more than this number?

24 **34**

Count on 10 more from each number. Show the card number and the number that is 10 more, for example, 24, 34.

What do you notice about each pair of numbers? (When you add 10 more, the tens number changes but the ones number stays the same.)

Discuss the students' observations.

Missing number

Materials

- number cards for 3, 13, 23, 33, 43, 53, 63, 73, 83, 93 and 103

Ask the students to sit in a circle. Place the set of number cards in the middle. Invite two students to stand.

Can you place these cards in order from smallest to largest?

After students have placed the cards in order, ask them to close their eyes. Remove one number card and rearrange the remaining cards. Ask the two students to look at the cards.

Which number is missing? How did you work it out?

Discuss the students' responses. Choose another two students and repeat the activity.

Arranging digits

Materials

- two sets of number cards from 0 to 9

Shuffle the two sets of number cards. Place them in a pile. Invite a student to take the top two cards, such as 4 and 5, and show the cards to the class.

Which 2-digit numbers can you make using these digits? (You can make 45 and 54.) What is the largest number you can make? How do you know that 54 is more than 45?

Discuss the students' understanding of place value.

Near enough

Materials

- 6 dice

Divide the class into three teams. Give each team two dice. Record a number between 40 and 66, such as 56. Invite a student from each team to roll their dice. Ask each student to use their dice to make a 2-digit number. Ask them to make a number that is as near as possible, but not greater than, the number recorded on the board. The team who is the closest wins a point for their team. Choose a new student from each team. Continue the activity until all students have had a turn.

Highest challenge

Materials

- two large dice

Ask the students to sit in a circle. Choose one student to stand behind another in the circle. Roll two dice into the middle of the circle. Ask the two students to call out the smallest number that can be made using the two numbers rolled. The first of the two students to call this out successfully moves on to stand behind the next student in the circle, and the other sits down or remains seated. Repeat the activity so all students in the circle have a turn.

Twenties tower

Materials

- 30 Multilink or Unifix cubes (10 cubes of 3 different colours)

Make two towers of 10 cubes:

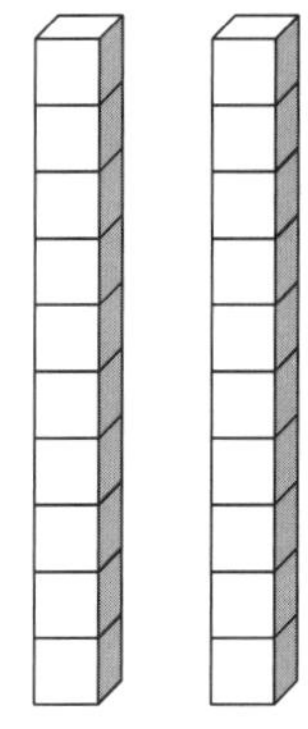

How many cubes are there? How many more cubes do I need to make 23 cubes altogether?

Make a third tower of 3 cubes.

23 is made of 10 and 10 and . . . . How many cubes do I need to take away from 23 to leave 20 cubes? How many cubes do I need to take away from 23 to leave 10 cubes?

Repeat the activity for 28 cubes, 21 cubes and 27 cubes.

Estimating cubes

Materials

- 90 Multilink or Unifix cubes

Show the students a pile of approximately 90 cubes.

How many cubes do you think are in this pile?

Choose individual students to count 10 cubes and connect them together. Encourage all students to revise their estimate after each 10 has been grouped.

What will be the best way to count the total number of cubes?

Count by tens to find the total. Repeat the activity using different numbers of cubes.

Ten-frames count

Materials

- ten-frames
- paper
- pencils

Show these ten-frames:

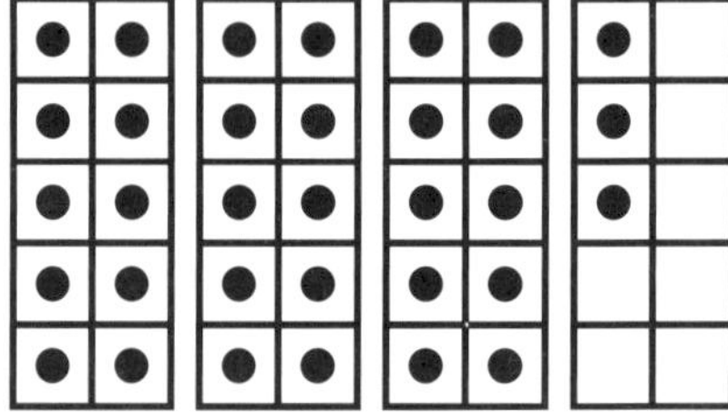

Ask the students to record the number of dots.

How many dots are there? How do you know that there are 33 dots? (I saw 10 and 10 and 10 and 3. I saw 30 and I counted 3 more.)

Repeat the activity with a new number. Discuss the students' responses.

More ones

Materials

- ten-frames

Briefly show these ten-frames, and then cover them:

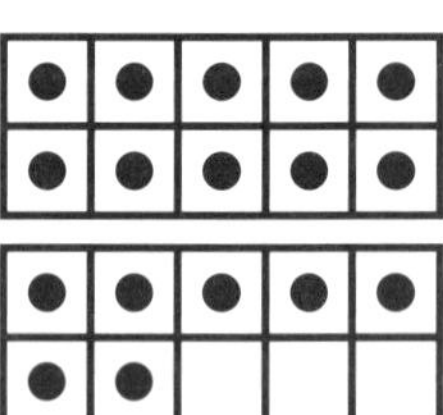

Did you see 17 dots? How many more dots did you see? How did you work it out? (20 take away 17 is 3. I counted from 17 to 20.)

Discuss counting back from 20 to 17 to verify the result. Repeat the activity for different numbers.

More tens

Materials

- ten-frames

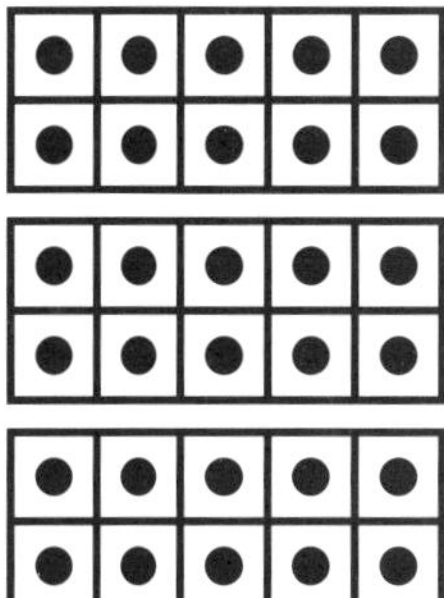

Briefly show 3 ten-frames, and then cover them:

Did you see 40 dots? How many were missing? (There are 30. 10 are missing. 1 ten-frame must be missing.)

Show another blank ten-frame. Discuss counting on from 30 to 40 to verify the result. Repeat the activity using different numbers of ten-frames.

Year 2

Imagining tens

Show this number line:

Tell the students that the line starts at 0 and goes up by tens.

Where does number 50 go? Where does number 80 go? What number is halfway between 60 and 70? Where does number 37 go? Is it closer to 30 or 40? What numbers are between 80 and 90? What number is 30 more than number 40?

Craft stick count

Materials

- a bucket
- 8 bundles of 10 craft sticks

Tell the students that there are 3 craft sticks in the bucket. Continue to place craft sticks in bundles of 10 in the bucket, asking students to keep the count.

Can you help me to count the number of craft sticks in the bucket? (13, 23, 33, 43 . . .)

When the students count to 83, repeatedly take bundles of 10 craft sticks from the bucket. *(83, 73, 63, 53 . . .)*

Ask the students to continue to count backwards by ten, off the decade. Repeat the activity, starting with different numbers of craft sticks.

Estimating counters

Materials

- 80 counters

Show the students approximately 80 counters.

Can you estimate how many counters are there altogether?

Ask the students to remember their estimate. Remove 10 counters from the collection.

Can you use these 10 counters to help you to estimate the total number of counters?

Remove 4 more groups of 10 counters. Encourage students to review their estimate after each 10 is counted. Ask the students to tell a partner their final estimate. Count all counters to determine the actual total.

Counting off the decade

Materials

- 5 ten-strips
- counters

Show this ten-strip:

How many dots can you see?

Add another ten-strip with ten dots. Continue to add ten-strips with ten dots and ask the students to count as you add.

How many dots are there now altogether? (4, 14, 24, 34. . .)

Variation

Begin with the total number of ten-strips and ask the students to count as each ten-strip is removed (34, 24, 14, 4. . .).

Counting craft sticks

Materials

- a bucket
- 8 bundles of 10 craft sticks

Tell the students that there are 52 craft sticks in the bucket. Continue to place bundles of 10 craft sticks in the bucket, asking students to keep the count.

Can you help me to count the number of craft sticks in the bucket? (62, 72, 82, 92 . . .)

When the students count to 132, repeatedly take bundles of 10 craft sticks from the bucket. Ask the students to continue to count backwards by ten off the decade.

(132, 122, 112, 102 . . .)

Repeat the activity, starting with different 2- and 3-digit numbers.

Block count

Materials

- a bucket
- Base 10 materials

Place these Base 10 materials in a bucket, one block at a time. Ask the students to keep a running total as each item is placed in the bucket.

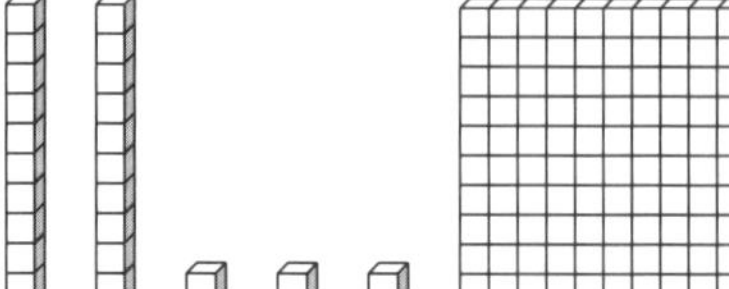

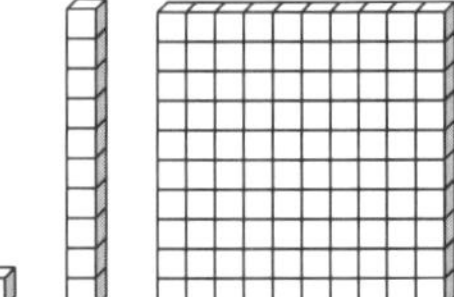

(1, 11, 21, 22, 23, 24, 124, 125, 135, 235 . . .)

How many tens are there now? How many ones?

Extension

Remove the blocks from the bucket one at a time. Ask the students to identify the number of blocks remaining.

Jumps on the decade

Show a number line from 200 to 300, with intervals of 10:

Invite students to count by tens from 200 to 300, forwards and backwards.

If I started at 200, how many jumps of 10 would it take to get to 300? If I started at 300, how many jumps of 10 would it take to get to 240? If I was at 240 and jumped forward three more jumps of 10, where would I land?

Discuss the students' strategies for obtaining answers.

Jumps off the decade

Show a number line from 32 to 132, with intervals of 10:

32 132

Invite students to count by tens from 32 to 132, forwards and backwards.

If I started at 32, how many jumps of 10 would it take to get to 132? If I started at 32, how many jumps of 10 would it take to get to 72? If I started at 52 and jumped forward four more jumps of 10, where would I land?

Discuss the students' strategies for obtaining answers.

Ten more and less

Materials

- number cards for 57, 43, 96, 25 and 108

Show the students one card at a time.

What is the number that is 10 more than this number? What is the number that is 10 less than this number?

Write the patterns formed around each number, for example, 47, 57, 67 . . . and 33, 43, 53. Use these patterns to count forwards and backwards by tens.

One hundred more or less

Materials

- number cards for 342, 117, 726, 103 and 691

Write these numbers on separate cards:

Show the students one card at a time.

What is the number that is 100 more than this number?
What is the number that is 100 less than this number?

Call for answers. On the board write the patterns formed around each number, e.g. 242, 342, 442 ...

100 more than 342 is 442. 100 less than 342 is 242. What do you notice about these numbers?

Two before and after

Materials

- a class set of calculators

Enter [1] [5] [3] on a calculator display.

What is the number that is 2 more than this number? What is the number that is 2 less than this number? How can we check the answers using the calculator?

Show the students how to enter [1] [5] [3] [−] [2] and [1] [5] [3] [+] [2] on the calculator to check. Repeat for other 3-digit numbers.

Extension

When students are ready, investigate 10 before and after and 100 before and after a chosen 3-digit number.

Showing triangles (1)

Show 10 triangles. Tell the students that they are going to use the triangles to help them to skip count by threes.

How many sides altogether can you see on these triangles?

Record the progressive total below each triangle.

3 6 9 12 15 18 21 24 27 30

How many triangles make a total of 21 sides? How many sides on 5 triangles? How many triangles would be needed to make 39 sides? What is the pattern that you notice here?

Encourage the students to explain how their answers were obtained.

How many more tens?

Materials

- ten-frames

Briefly show three ten-frames with 10 dots each, and then cover them:

Did you see 50 dots? How many were missing? (20. . . 2 tens.)

Show the extra two ten-frames. Discuss counting on from the 30 to the 50 to verify the result. Repeat using a different number of ten-frames.

How many more ones?

Materials

- ten-frames

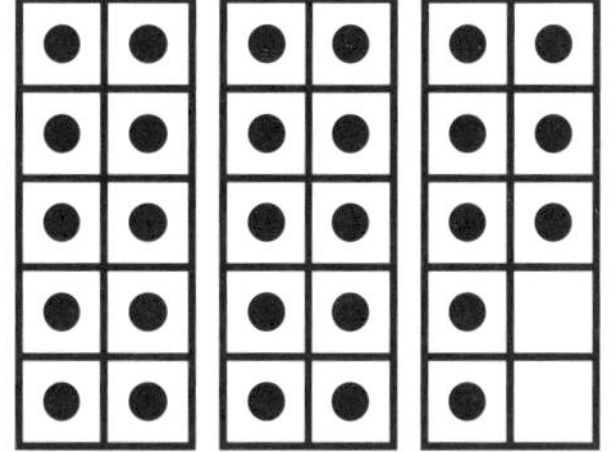

Briefly show three ten-frames, and then cover them:

Did you see 28 dots? How many more dots did you see? How did you work it out? (I counted from 28 to 30. 30 take away 2 is 28.)

Show the ten-frames again. Discuss counting back from 30 to 28 to verify the result. Repeat using a different number of ten-frames.

Two-digit breakdown

Materials

- Unifix or Multilink cubes or Base 10 materials

Show this number: **58**

How many tens in 58? (There are 5 tens.)

How many ones in 58? (There are 58 ones. There are 5 tens and 8 ones.)

Can you picture 58 ones? Can you picture 5 tens and 8 ones? What do they look like?

Use materials to demonstrate the different representations of 58, so that students can develop more complete mental pictures and understandings.

One before and after

Materials

- a class set of calculators

Enter [7] and [5] on the calculator display.

What is the number that is 1 more than this number? What is the number that is 1 less than this number? How can we check the answers using the calculator?

Show the students how to enter [7] [5] [–] [1] and [7] [5] [+] [1] on the calculator to check. Repeat for other 2-digit numbers.

What is the number?

Materials

- a counter
- a hundreds chart

Write the numbers on the chart as shown. Tell the students that some of the numbers are missing from the hundreds chart.

Can you imagine the missing numbers?

Place a counter or cover on a square.

What is this number on the hundreds chart? How do you know?

Repeat this activity often by changing the location of the counter on the hundreds chart.

1	2	3	4	5	6	7	8	9	10
11									
21									
31									
41			●						
51									
61									
71									
81									
91									

Around the number

Materials

- a counter
- a blank hundreds chart

Show a blank hundreds chart. Write the numbers 1 to 10 across the top row of the chart. Tell the students that some of the numbers are missing from the chart.

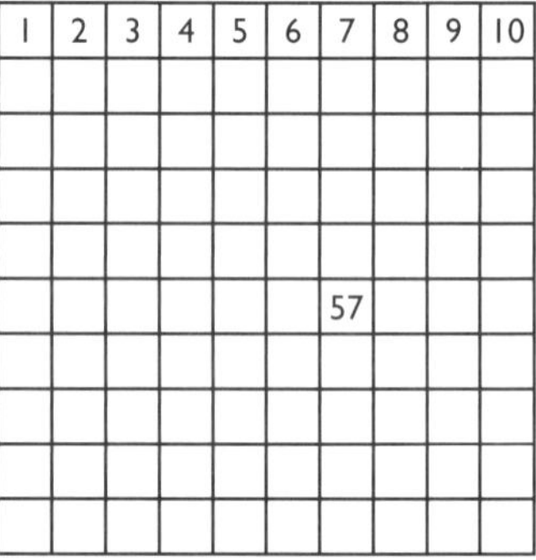

Can you imagine the missing numbers?

Place a counter on a square, for example, 57.

What is this number on the hundreds chart? How do you know? Which numbers are in the squares around this number? (46, 47, 48, 58, 68, 67, 66, 56)

Repeat this activity often by changing the location of the counter on the hundreds chart.

Hundreds chart puzzle

Show this puzzle with the numbers shown:

	63	64		
		74		
		84		

Can you use a pattern to find the missing numbers? How did you work out your answers? (The number underneath is 10 more, the number to the left is 1 less, the number to the right is 1 more – 53, 62, 65, 73, 83, 85, 86, 95.)

Repeat the activity for different shaped puzzles.

Hundreds chart path

Show this number path:

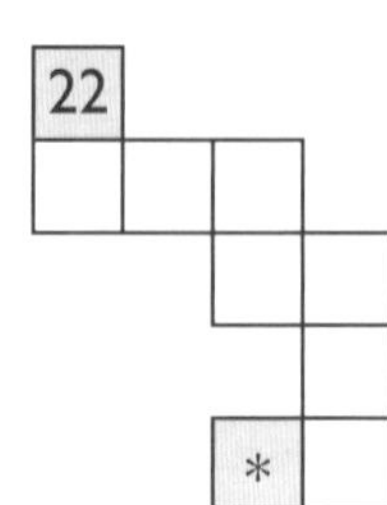

Tell the students that the path is from a hundreds chart.

Can you work out the numbers on the path? What did you think of when working out the numbers? (The number underneath is 10 more, the number to the left is 1 less, the number to the right is 1 more – 32, 33, 34, 44, 45, 55, 65, 64.)

Repeat the activity for other different shaped paths.

Hundreds chart jigsaw

Materials

- a paper hundreds chart for each student
- scissors

Give each student a hundreds chart. Ask them to cut their hundreds chart into six different pieces, following the grid lines. Ask the students to swap their hundreds chart with a partner and attempt to put the chart back together.

What did you use to put the jigsaw back together? Can you use number patterns to put the jigsaw back together?

Comparing sticks

Materials

- 7 bundles of 10 craft sticks
- 15 single craft sticks

Show the students a collection of 4 bundles of 10 craft sticks and 6 single craft sticks, and another collection of 3 bundles of 10 craft sticks and 9 single craft sticks.

Which number is greater? How do you know? (The first collection is larger. You need to add more to 39 to get 46. When I count, I say 39 before I say 46. 46 has more tens.) How can you check to make sure?

Repeat the activity for other numbers.

Ordering by tens

Materials

- blank number cards

Show these numbers on separate cards:

71 **41** **91** **51** **61**

Choose five students to choose one number card each.

Can you order these numbers from lowest to highest? Which digit is important in deciding how to order these numbers? Why? (They all have different numbers of tens. You can order the numbers by looking at the tens digit. . . 41, 51, 61, 71, 91.)

What is the second-highest number in the sequence? (71) How do you know?

Ordering by tens and ones

Materials

- blank number cards

Show these number on separate cards:

42 **75** **49** **27** **72**

Choose five students to choose one number card each.

Can you order these numbers from lowest to highest? (27, 42, 49, 72, 75) How do you know that 75 is the highest number? (75 is higher than 72 because 75 has 5 ones and 72 has 2 ones.)

What is the lowest number in the sequence? How do you know? (27 is the lowest number because it has only 2 tens.)

Two-digit wipe out

Materials

- a class set of calculators

Enter [7] [2] on the calculator display.

How could you make the 7 disappear? (Subtract 72. Subtract 70. Subtract 7.)

Why did you subtract 70 and not 7? (It is because the 7 stands for 70, not 7.)

Ask these questions and discuss the students' responses.

How could you change the 2 in 72 to 0? (Subtract 2.) How could you change the 7 in 72 to a one? (Subtract 60.) How could you change 72 to 172? (Add 100.)

101–200 chart

Materials

- a counter
- a blank hundreds chart

Show a blank hundreds chart. Write the numbers on the chart as shown. Tell the students that some of the numbers are missing from the chart.

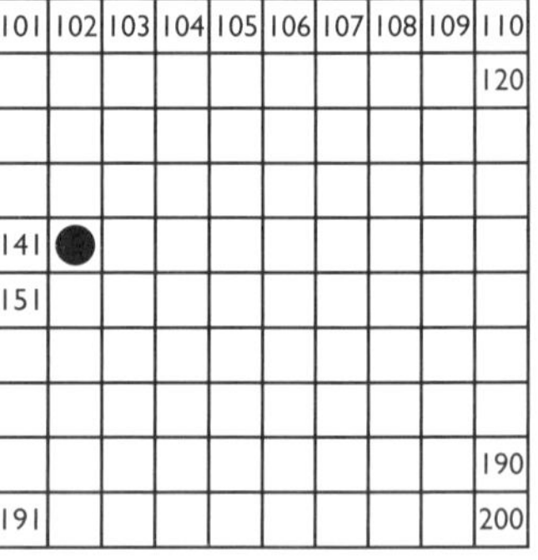

Can you imagine the missing numbers?

Place a counter on a blank square.

What is this number on the hundreds chart? (142) How do you know?

Repeat this activity often by changing the location of the counter on the hundreds chart.

Imagining numbers

Show this number line:

0 50 100 150

Where does number 60 go? Where does number 110 go? Which 3 numbers are halfway between 0 and 50? Where does number 79 go? Is it closer to 50 or 100? Which numbers are between 50 and 100? Which number is 50 more than number 100? Which number is 50 more than 150?

Discuss the students' strategies for obtaining answers.

What is my number?

Show this number line:

0 150

Choose a secret number between 0 and 150. Invite students to guess the number.

I am thinking of a number between 0 and 150. Can you guess my number?

Respond to the students' guesses.

My number is less than/greater than . . .

Mark each guess on the number line. Continue until the chosen number is determined.

Repeat the activity using other number lines.

Where is the number?

Show a number line from 60 to 200, with intervals of 10:

How can you work out where 180 would be? Is 180 closer to 60 or 200? How do you know?

Mark the position on the number line. Repeat for other numbers.

Variation

Point to another position on the number line.

Which number am I pointing to? How did you work it out?

Higher or lower

250

100

Show this vertical number line:

Choose a secret number in this range. Invite students to guess the number.

I am thinking of a number between 100 and 250. Can you guess my number?

Respond to the students' guesses.

My number is higher/lower than . . .

Mark each guess on the number line. Continue until the chosen number is determined. Repeat the activity using other number lines.

Order by ones

Materials

- blank number cards

Write these numbers on separate cards:

57	53	59	51	55

Choose five students to choose one number card each.

What is the same about all these cards? (They all have 5 tens.) Can you order these numbers from lowest to highest? (51, 53, 55, 57, 59) Which digit is important in deciding how to order these numbers? Why? (You can order the number by the size of the ones.)

Ordering numbers

Materials

- string
- pencils
- cardboard
- pegs

Give each student a piece of cardboard. Ask them to write a number between 100 and 200 on it.

Make an open number line by tying a length of string across the room. Place number 100 at one end and number 200 at the other. Invite students to place their number card in the appropriate position on the line.

Can you read your number to me? Is your number closer to 100 or 200? What is the number before/after your number?

Greater than or less than

Show these number pairs: **34 and 37, 67 and 39, 85 and 80, 112 and 97**

Consider the numbers one pair at a time.

Which of these two numbers are greater? (112 is greater because it has 11 tens. . . 97 has 9 tens. . . 67 is bigger as it has 6 tens. . .) Which is lesser? Which digit is important in deciding whether the number is greater?

Discuss the relative sizes of the numbers in terms of their place value.

Three-digit breakdown

Materials

- Unifix or Multilink cubes or Base 10 materials

Show this number: **237**

How many hundreds in 237? (There are 2 hundreds.) How many tens in 237? (There are 23 tens.) How many ones in 237? (You could make the number with 237 cubes.) Can you picture 237 ones? Can you picture 23 tens and 7 ones? Can you picture 2 hundreds, 3 tens and 7 ones? Can you picture 2 hundreds and 37 ones? What do they look like?

Use materials to demonstrate the different representations of 237, so that students can develop more complete mental pictures and understandings.

Arranging digits

Show these numbers: **3, 6, 5**

Which 3-digit numbers can you make using these digits? (635, 356, 653, 365, 563, 536) What is the largest number you can make with these three digits? (653) Why did the largest number start with 6?

Discuss the students' understanding of place value.

Largest and smallest

Materials

- number cards from 0 to 9

Show the students a set of number cards from 0 to 9. Choose three students to choose a card from the deck.

How could you arrange the number cards to make the largest possible number? Why did you arrange the cards that way? (You put the largest number in the hundreds place.) What is the smallest possible number? Why did you place the smallest number in the hundreds place? Can zero be used to begin your number? (No. Zero holds the place in the middle or at the end of the number only.)

Repeat the activity with another group of three students.

Digit show

Materials

- two sets of number cards from 0 to 9
- two 3-column place value charts

Divide the class into two teams. Shuffle the number cards. Explain to the students that the aim of the activity is to make the largest 3-digit number possible.

Rules for digit show

1. A student from each team randomly chooses a number card from the deck.
2. Both students place their cards on their team's place value chart, in the column of their choice.
3. After three choices, the team with the highest number on their place value chart is the winner.

TEAM A

H	T	O
8		

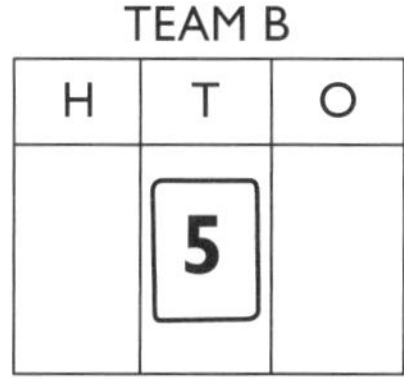

Representing numbers

Materials

- a class set of calculators

How can you make the calculator display 137 using the [+] *key? (130 + 7, 100 + 37)*

Discuss the students' responses. Emphasise the answer as

[1] [0] [0] [+] [3] [0] [+] [7].

Can we show other 3-digit numbers like this? How could we show 431?

Invite students to enter their responses into the calculator to check solutions.

Three-digit wipe out

Materials

- a class set of calculators

Enter the number [1] [6] [5] on the calculator.

How could you change the 6 on the show to 0? Why did you subtract 60 and not 6? (It is because the 6 stands for 60 not 6.)

How could you change the 5 in 165 to a zero? How could you make the 1 disappear? How could you change the 1 in 165 to a seven?

Discuss the students' responses.

Three-digit change

Materials

- a class set of calculators

Enter [1] [3] [5] on the calculator.

How could you change the 3 on the show to a 7? Why did you add 40 and not 4? (You must add 4 tens, not 4 ones.)

How could you change the 1 in 135 to a 7? How could you change the 5 in 135 to a 9? How could you change the 3 in 135 to a 0?

Discuss the students' responses.

Addition and subtraction

Year F

Describing operations

Materials

- 10 cardboard fish cut-outs

On Monday, Gavin and Catherine went fishing. Gavin caught 4 fish and Catherine caught 5 fish. How many fish did they catch altogether? How did you work it out? (You add the 4 fish to the 5 fish. That is 9 fish.)

Invite students to model the problem using the fish cut-outs. Encourage them to describe the action of addition.

Gavin and Catherine gave 5 of the fish to their neighbour. How many fish do they have left? How did you work it out? (You start with 9 and take 5 away. That is 4 left.)

Invite students to model the problem using the fish cut-outs. Encourage them to describe the action of subtraction.

Add one more

Can you show me 6 fingers?

Ask the students to then put their hands behind their back. Tell them to put up one more finger without looking at their hands.

How many fingers are you showing now?

Encourage students to state their answer without looking at their fingers. Observe whether students need to count their fingers or whether they can instantly recognise the total. Repeat the activity, asking students to show one more than 8 fingers, 2 more than 4 fingers and 3 more than 5 fingers.

Take one away

Can you show me 8 fingers?

Ask the students to then put their hands behind their back. Tell them to put down one of the fingers without looking at their hands.

How many fingers are you showing now?

Encourage students to state their answer without looking at their fingers. Observe whether students need to count their fingers or whether they can instantly recognise the total. Repeat the activity, asking students to put down one from 6 fingers, 2 from 5 fingers and 2 from 2 fingers.

Show me

Ask the students to show a particular number of fingers. Encourage them to show the fingers instantly, without having to count them one at a time.

Can you show me 4 fingers? Can you show me 8 fingers?

Encourage the students to count on from a particular number of fingers.

Can you show me one more than 4 fingers? Can you show me 3 fingers more than 1 finger? Can you show me 1 finger less than 7 fingers?

Discuss the students' responses.

Making groups

Materials

- counters

Give 5 counters to each student.

Can you make a group of 3 counters and a group of 2 counters?

Show 3 counters and 2 counters.

How many counters have you used altogether? Can you use the 5 counters to make two different groups?

Allow students some time to investigate the different combinations that can be made using 5 counters. Record all possible combinations on the board using Showings. Repeat the activity using 8 counters.

Counter hands

Materials

- counters

Place 7 counters in one hand and 2 counters in the other. Open one hand.

I have 7 counters in this hand.

Close this hand. Open the second hand to show 2 counters.

I have 2 counters in this hand. How many counters do I have altogether? How did you work it out?

Encourage students to explain their strategies. Repeat the activity for other combinations to at least 10.

More counter hands

Materials

- counters

Place 5 counters in one hand and 2 counters in the other. Open the hand with 5 counters.

I have 5 counters in this hand.

Close this hand.

I have 7 counters altogether in my hands. How many counters do I have in my other hand if there are 7 counters altogether? How did you work it out?

Encourage students to explain their strategies. Repeat the activity for other combinations to at least 10.

Finger count

Cynthia had 8 balloons. She gave 5 of the balloons to her friend. How many balloons does Cynthia have left? Can you use your fingers to work it out?

Choose students to demonstrate how they could solve the problem by counting on their fingers. Encourage students to use their fingers to solve another problem.

Stephanie had 7 balloons but 2 of the balloons popped. How many balloons does Stephanie have left? How could you use your fingers to work it out?

Observe the students' strategies.

Find a partner

Materials

- Multilink or Unifix cubes

Invite all students to choose either 1, 2, 3, 4 or 5 cubes. Ask those with more than one cube to make a tower. Ask them to find a place in the room and hold their tower for all to see.

Can you join with another person to make 6 cubes?

Tell the students that they must use all their cubes. Ask all students without a partner to sit. Ask those who are standing to take back their original number of cubes. Repeat the activity with the students who are still standing. Ask the students to join with a partner to make a particular total until only two students remain standing.

Sticker problem

Show this diagram:

Bala has collected 8 stickers. Mia has 5 stickers. Mia wants to collect the same number of stickers as Bala. How many more stickers does Mia need? How did you work it out? How could you use the diagram to work it out?

Encourage the students to demonstrate and explain their strategies. Repeat the activity.

How many more stickers does Bala have than Mia?

Counter containers

Materials

- 10 counters
- 2 small buckets or containers

Ask the students to sit in a circle. Place two small buckets or containers upside down on the floor. Place 5 counters in the first container and 3 counters in the second container.

Can you imagine the counters in the containers? How many counters are there altogether? How did you work it out?

Encourage students to explain their strategies. Repeat the activity for these combinations: 4 and 5, 8 and 2, and 9 and 1.

In a box

Materials

- 12 cubes
- 2 small boxes

Show the students two small boxes. Place 6 cubes in one box and 3 cubes in the other. Invite two students to count how many cubes are in each box.

Can you tell the class how many cubes are in each box? How many cubes are there altogether? How did you work it out?

Encourage students to explain their strategies. Repeat the activity for these combinations: 2 and 5, 7 and 2, and 10 and 1.

Extension

Extend the combinations beyond 10, for example, place 7 cubes in one box and 5 in the other.

Tommy the toad

Show this number line:

1 2 3 4 5 6 7 8 9 10

Tommy the toad lived at zero on the number line. He wanted to visit his friend at number 10. In the morning, he jumped forward 4 jumps.

Show Tommy's jumps on the number line as follows:

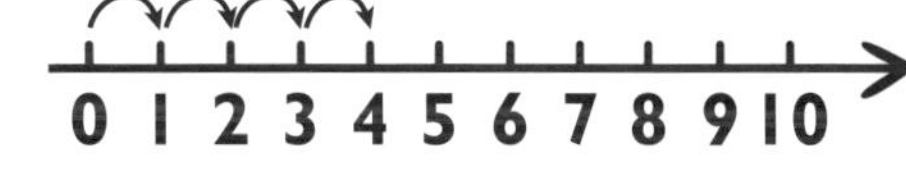

How many jumps does Tommy have to make in the afternoon to reach number 10?

Ask the students to demonstrate their thinking.

Extension

Repeat the activity for different numbers of morning and afternoon jumps.

On the bus

James is a school bus driver. At the first bus stop, 5 children got on the bus. At the next bus stop, 3 children got on the bus. How many children are on the bus? How did you work it out?

Choose students to demonstrate and explain their strategies.

Some more children got on the bus at the next bus stop. There are now 10 children on the bus. How many children got on at the third bus stop? How did you work it out?

Discuss the students' thinking.

More stickers

Jasmine has collected 4 stickers. Saki has 5 more stickers than Jasmine. Can you show a picture to work out the answer to the problem?

Encourage students to draw a picture to represent the problem. Invite some students to share their drawings.

How many stickers does Saki have? How did you work it out?

Encourage students to demonstrate and explain their strategies.

Highest total

Materials

- 4 dot dice

Divide the class into two teams. Give each team two dot dice. Choose a student from each team. In turn, ask the students to roll their two dice.

What is the total of the numbers shown on the dice? Which team has rolled the highest total?

The team with the highest total wins a point. Choose a different student from each team to roll the dice. Repeat the activity until all students have had a turn. Throughout the activity, choose some students to explain how they determined the total.

Add and compare

Materials

- ten-frames with varying amounts of dots
- counters

Ask the students to sit in two teams. Place a set of ten-frames face down in one pile. Ask the first team to turn over two cards and state the total number of dots. Then ask the second team to do the same.

Which team has more dots?

The team with the greater number of dots wins a point. Return the cards to the pile and shuffle the pile of cards. Repeat the activity until a team wins 10 points. Include random and non-random arrangements of dots on the ten-frames.

Ten-frame doubles

Materials

- ten-frames

Show two blank ten-frames. Place 4 counters in each ten-frame:

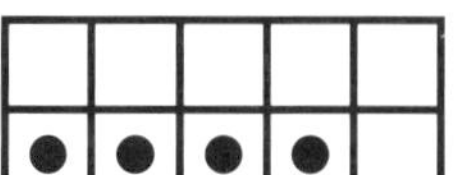

How many dots can you see? How do you know that there are that many? (I saw 4 and 4. That makes 8.)

Discuss the students' strategies. Repeat the activity doubling other numbers from 0 to 10.

Frame doubles plus one

Materials

- ten-frames

Show two blank ten-frames. Place 4 counters in one ten-frame and 5 in the other.

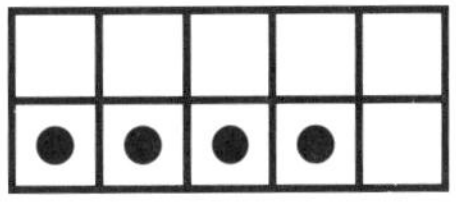

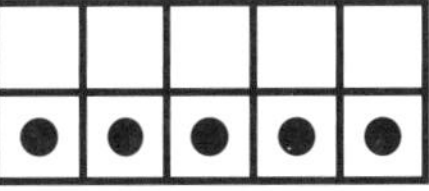

How many dots can you see? How did you work it out?

Discuss the strategy of doubles plus one. Repeat the activity for other numbers from 0 to 10.

Taking off

Materials

- ten-frames
- counters

Show a blank ten-frame. Arrange 9 counters in the ten-frame.

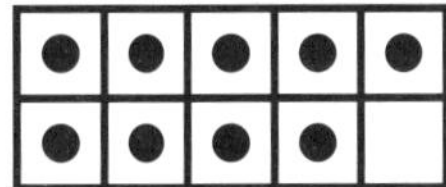

How many counters can you see? How many squares are empty? How did you work it out?

Discuss the students' strategies.

Can you imagine what it would look like if someone took 3 counters off the frame? How many counters would we have?

Choose students to explain their thinking to the rest of the class. Repeat the activity, subtracting other numbers from numbers 1 to 10.

Make it the same

Materials

- ten-frames
- counters

Show two blank ten-frames. Place 5 counters in 1 ten-frame and 7 in the other.

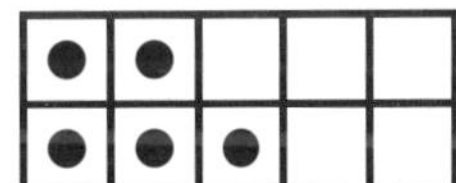

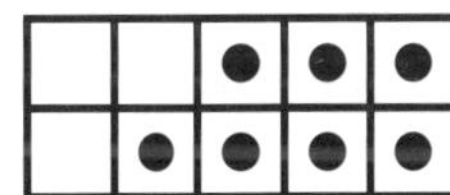

How many counters can you see in each frame? Can you imagine how many counters are needed to make each group the same?

Discuss the students' strategies. Discuss the idea of comparison. Repeat the activity for other numbers from 0 to 10.

Three dice add

Materials

- 3 dot dice

Invite three students to stand. Give each student a dot dice. Ask the three students to roll their dice at the same time.

What is the total of the 3 dice?

The first student to call the answer remains standing. The other two students sit, and two new competitors are chosen. Throughout the activity, ask the winning student to explain how they quickly determined the answer. Continue the game until all students have had a turn.

Subtraction line-up

Materials

- a dice

Divide the class into two teams. Ask each team to line up facing the board. Choose one student from each team to write the total number of students in their team on the board. Ask a student from the first team to roll the dice, read the number shown on the dice and ask that number of students from the team to sit. The second team repeats this process.

How many students in your team are left standing?

Continue the activity until all students from a team are seated. The correct number must be rolled before the last student can sit.

Two numbers

Show this diagram:

Point to each group of dots in turn.

How many dots are there in this group of dots?

Tell the students that you are going to imagine putting together two of the dot groups.

When I add two of the dot groups together, they make 8 dots. Which two groups are they? How did you work it out?

Count the dots with the students to check. Repeat the activity for totals of 10 dots, 9 dots and 7 dots.

Estimating addition

Materials

- 17 red counters
- 8 blue counters

Show the students a pile of 17 red counters and 8 blue counters.

Can you estimate the number of red counters? What about the number of blue counters?

Accept all estimates.

Can you estimate the total number of counters?

Count the actual number of red counters by placing them into a pile of 10 counters and a pile of 7 counters. Count the number of blue counters. Record the number of red and blue counters.

How can we work out the total number of counters?

Share and discuss the students' strategies.

Year 1

Subitise and add

Materials

- dot dice cards

Show the students a dot card:

Can you quickly tell me how many dots you see?

Encourage students to state the number of dots instantly (without stopping to count each dot).

Can you picture the number of dots in your mind?

Cover the card and show the students a second dot card:

How many dots do you see? How many dots are there altogether on the two cards? How did you work it out? (I put up 5 fingers and 4 fingers and counted my fingers. I know that double 5 is 10, so 5 plus 4 is 9.) Is there a quicker way of working out the total number of dots? (I counted 5, 6, 7, 8.)

Discuss and share the students' strategies.

Subitise and subtract

Materials

- dot dice cards

Show the students a dot card:

Can you quickly tell me how many dots you see?

Encourage students to state the number of dots instantly (without stopping to count each dot).

Can you picture the number of dots in your mind?

Remove the card.

How many dots will there be if I take 2 dots away? How did you work it out? (I put up 6 fingers and then put 2 fingers down. I imagined that 2 dots went away.) Is there a quicker way of working out how many dots are left? (I counted 6, 5, 4.)

Discuss and share the students' strategies.

Ten-frames race

Materials

- a dot dice

Divide the class into two teams with a blank ten-frame for each team:

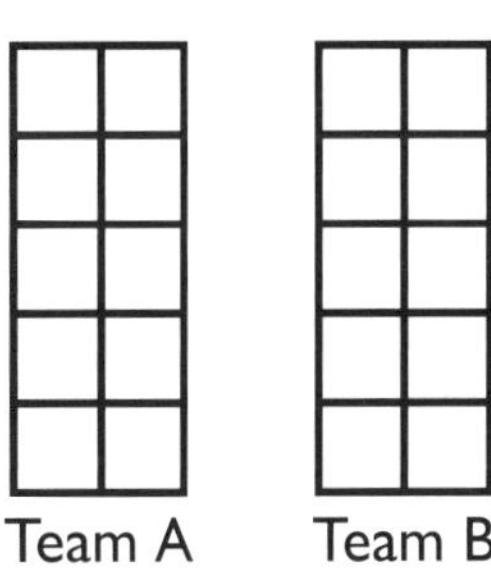

Rules for ten-frames race

1. The aim of the game is to be the first team to cross off each square of their ten-frame.
2. One student in each team rolls the dice and crosses off the corresponding number of squares on their ten-frame.

3. Each team must then roll the exact number of remaining blank squares to win (e.g. if a team has filled 8 squares and rolls a 3, the team must wait for another turn to try and roll a 2).

Using fingers

Mum gave Sasha 12 grapes to eat, but 5 grapes fell on the ground and could not be eaten. How many grapes does Sasha have left to eat? How could you use your fingers to work it out?

Choose students to demonstrate their solutions. Encourage students to use their fingers to solve the next problem.

Abdul and Henry are playing with their toy cars. Abdul has 7 cars and Henry has 8 cars. How many toy cars do Abdul and Henry have altogether? How could you use your fingers to work it out?

Observe the students' finger-counting strategies.

Jump to 20

Materials

- 2 dice
- chalk

Divide the class into two teams. Nominate one student from each team as the 'jumper'. Show a large number line or track from 0 to 20 on the floor:

Rules for jump to 20

1. Each team in turn rolls the two dice.
2. Team members subtract the smaller number from the larger number.
3. The team 'jumper' moves forward that number of places on the number line.
4. The winner is the team whose 'jumper' reaches 20 first.

How much further did the winning team jump than the other team? How did you work it out?

Bucket teens

Materials

- craft sticks
- a bucket

Place a bundle of 10 craft sticks in the bucket. Tell the students that there are 10 craft sticks in the bucket. One at a time, place bundles of 5 more craft sticks in the bucket.

How many sticks are now in the bucket altogether? How many bundles of 10 sticks are in the bucket? How many sticks do I need to put in to make 20 sticks altogether?

Discuss the students' strategies. Repeat the activity for other teen numbers. Develop the language of ten and five, ten and six and so on during this activity.

Think of a number

Show this number line: 0 1 2 3 4 5 6 7 8 9 10 11 12 13 14 15 16 17 18 19 20

I am thinking of a number. When I add 2 to the number the answer is 8. What is the number? How did you work it out?

Discuss the students' strategies. Invite students to present similar problems for others to solve.

Line jumps

Show this number line:

Roo wanted to work out 4 plus 7 by using a number line. Roo started at zero and jumped forward 4 numbers. Then Roo jumped forward another 7 numbers. Which number did Roo land on? How did Roo work it out?

Encourage students to explain their thinking. Invite a student to show the jumps on the number line. Repeat the activity for 5 + 7, 10 + 2 and 3 + 9.

Toad subtraction

Show this number line:

Toad wanted to work out 11 minus 6 by using a number line. Toad started at 11 and jumped backwards 6 numbers. Which number did Toad land on? How did Toad work it out?

Encourage students to explain their thinking. Invite a student to show the jumps on the number line. Repeat the activity for: 10 – 5, 9 – 5, 12 – 8.

Counter combinations

Materials

- counters

Give 10 counters to each student.

Can you make a pile of 7 counters and a pile of 3 counters?

Show this number sentence: **7 + 3 = 10**

How many counters have you used altogether? Can you use the 10 counters to make 2 different piles?

Allow students some time to investigate the different combinations that can be made using 10 counters. Ask the students to record all possible combinations as number sentences.

Extension

Repeat the activity for number facts to 20 or more.

Combinations for 10

Show these numbers: **2, 9, 6, 8, 3**

Can you see two numbers that add together to make 10? What are they?

Observe whether students can instantly recall number facts for 10.

What two numbers add together to make 1 more than 10? How do you know?

Discuss the students' strategies. Repeat the activity often using different 1-digit numbers.

Make 10 bingo

Materials

- a 4 × 1 grid for each student
- a set of number cards from 1 to 9

Give each student a 4 × 1 grid. Ask the students to choose numbers between 1 and 9 and record one in each square, for example:

Choose a number card at random. Show the card to the students.

Have you got a number that when added to this number makes 10?

Ask the students to cross off the desired number if they have recorded it in their grid. Continue to play until a student has crossed off all numbers.

Variation

Vary the game to consolidate other addition and subtraction facts.

Counter cover-up

Materials

- 12 counters
- a cloth

Invite students to sit in a circle. Place 12 counters in a line on the floor. Ask the students to count the counters, and then close their eyes. Place the cloth over four of the counters.

How many counters have I hidden under the cloth? How did you work it out?

Discuss and model the strategies that students use. Repeat the activity, covering a different number of counters each time.

Hidden counters

Materials

- 13 counters
- a bucket

Invite the students to sit in a circle. Place a bucket upside down in the middle.
Place 13 counters on top of the bucket. Invite a student to count the number of counters.
Ask the students to close their eyes. Place 9 of the counters under the bucket.

How many counters have I placed under the bucket? How did you work it out? (I saw 4 counters on the bucket and I counted on my fingers to 13. I took 4 away from 13 to work it out.)

Share and discuss the students' strategies. Repeat the activity, varying the number of counters placed under the bucket.

Bucket remainder

Materials

- 20 Multilink or Unifix cubes
- a bucket

Show the students a bucket and a collection of blocks. Tell the students that you are going to drop the blocks into the bucket one at a time.

When I stop, can you tell me how many blocks are in the bucket?

Drop the 20 blocks into the bucket, one at a time.

If I took 6 blocks out of the bucket, how many blocks will be left in the bucket? How did you work it out? (There were 20 to start. I counted back 19, 18, 17 to 14.)

Discuss the students' strategies. Repeat the activity often. Ask the students to determine the number of blocks in the bucket when different numbers of blocks are taken away.

Bucket difference

Materials

- craft sticks
- a bucket

Show the students a bucket. Count 20 craft sticks one by one into the bucket. Tell the students that there are 20 craft sticks in the bucket, but you only need 15.

How many craft sticks must I take away to get 15? How did you work it out? (I counted 19, 18, 17, 16, 15 . . . on my fingers, 15 and 5 more is 20.)

Discuss the students' strategies. Repeat the activity for other numbers less than 20.

How many added?

Materials

- counters

Show the students 5 counters. Ask the students to close their eyes. Show another 12 counters. Ask the students to look at the new collection.

How many more counters did I add? How did you work it out? (There are 17 counters now; before there were 5 so17 take away 5 is 12. I counted 5 counters, and then I counted 12 more.)

Ask the students to explain or demonstrate their strategies. Repeat this activity often, adding a different number of counters each time.

How many did I take?

Materials

- counters

Show the students 15 counters. Ask the students to close their eyes. Remove 7 counters. Ask the students to look at the remaining counters.

How many counters did I take? How did you work it out? (There is 8 left. . . 9, 10, 11, 12, 13, 14, 15. You must have taken 7.)

Discuss the students' strategies. Repeat the activity, removing a different number of counters each time.

Comparing lengths

Materials

- Multilink or Unifix cubes

Marlee and Jayant were measuring the length of their shoes with small cubes. Marlee's shoe was 12 cubes long and Jayant's shoe was 16 cubes long. How many more cubes did Jayant use than Marlee? How did you work out this problem? (I started at 12 and I counted 4 more to get to 16.)

Discuss the students' strategies. Invite a student to count 12 cubes and make a tower. Invite another student to count 16 cubes and make a tower. Place the towers side by side to compare. Encourage students to describe how they could use the towers to find the difference between 12 and 16.

Twenty-frame race

Materials

- ten-frames
- a dice

Divide the class into two teams with two ten-frames for each team:

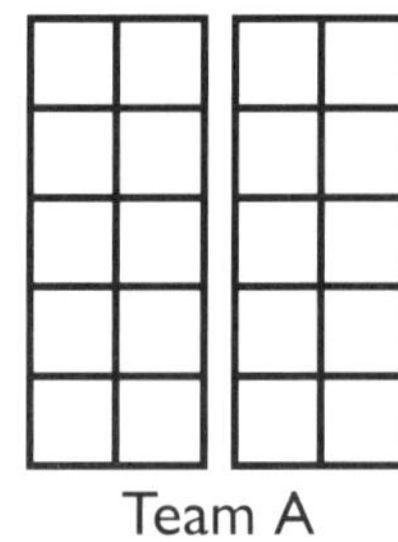
Team A

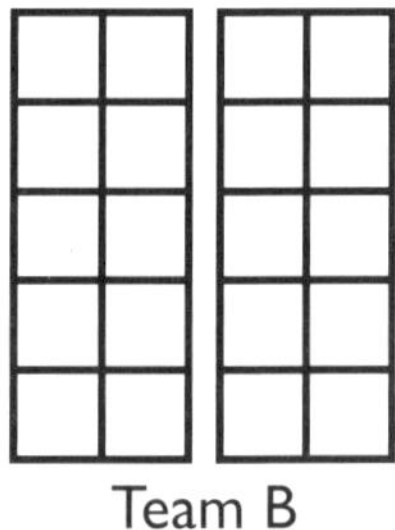
Team B

Rules for twenty-frame race

1. The aim of the game is to be the first team to fill each square of their ten-frames.
2. One member of each team rolls the dice and adds the corresponding number of dots in the squares on their team's ten-frames.
3. The team must roll the exact number needed to win (e.g. if a team has filled 16 squares and rolls a 5, the team must wait for another turn).

Strips to 20

Materials

- ten-strips

Show a ten-strip with 10 dots and a ten-strip with 8 dots:

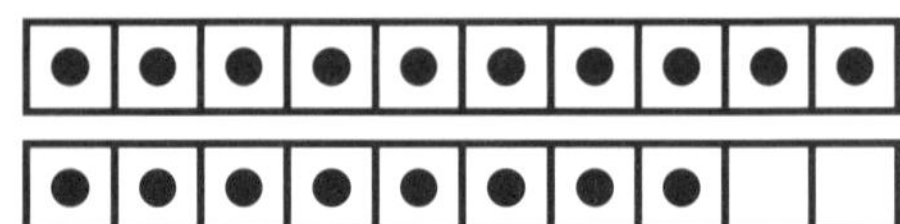

How many dots can you see? Can you work out the number of dots without counting by ones? (I can see 20 take away 2. There is 10 and 8 more.)

Discuss the students' strategies for counting the dots quickly.

How many more dots do I need to make 20 dots? How do you know?

Encourage students to explain their strategies. Repeat the activity for other numbers in the range of 10 to 20.

Bridge from nine

Materials

- ten-frames

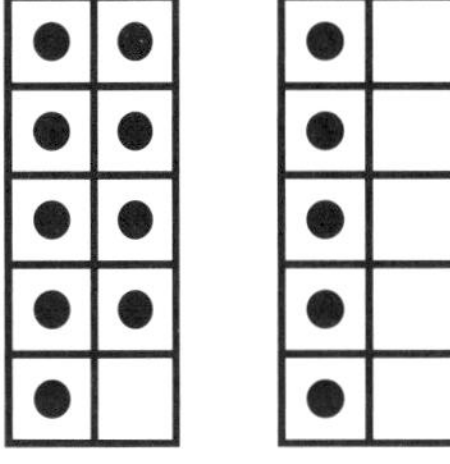

Show one ten-frame with 9 dots and another ten-frame with 5 dots:

How many dots did you see altogether? How do you know? (I imagined that 1 dot from 5 went over to the 9 to make 10 and 4 more. I saw 9 and 5. I put up 9 fingers and counted 10, 11, 12, 13, 14.)

Ask the students to explain their strategies. Show the ten-frames again to verify students' answers. Repeat the activity using other ten-frames.

Bridge from eight

Materials

- ten-frames

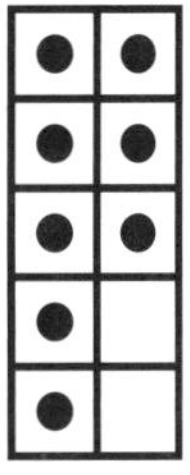

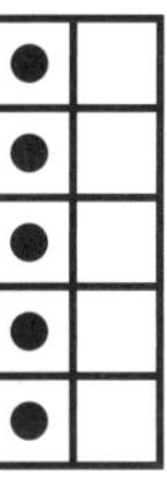

Show a ten-frame for the number 8 briefly:

How many dots did you see?

Ask the students to remember this ten-frame. Remove or cover the ten-frame for 8 and show or show the ten-frame for number 5 briefly:

How many dots are there if you add these dots to the 8 dots? How did you work it out? (You give 2 to the 8 and that makes 10. There is 3 left. 11, 12 . . . 13 altogether.)

Discuss the students' strategies.

Count back

Materials

- ten-frames

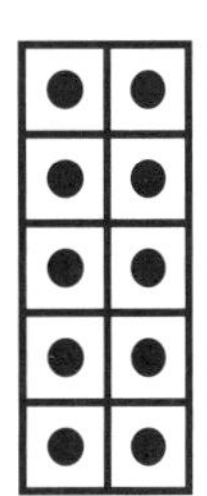

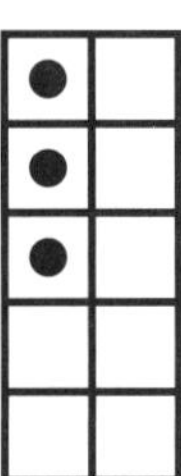

Briefly show one ten-frame with 10 dots and another with 3 dots:

How many dots did you see altogether?

Ask the students to remember this number. Remove the ten-frames. Show a ten-frame with 8 dots.

How many dots do you see now? How many dots did I take away? How did you work it out?

Discuss the students' strategies.

What is the difference?

Materials

- 2 dice
- 90 Unifix cubes

Invite the students to sit in two teams. A student from each team rolls the two dice in turn.

They state the total rolled and make a tower using that number of Unifix cubes. Compare the towers made by each team to determine the difference.

Whose tower has the most cubes? How many more cubes does this team have? What is the difference between these two towers?

The team with the larger tower keeps the number of cubes that is the difference. Repeat this process until one team collects a total of 30 cubes.

Bridge from 16

Show a number line from 12 to 24:

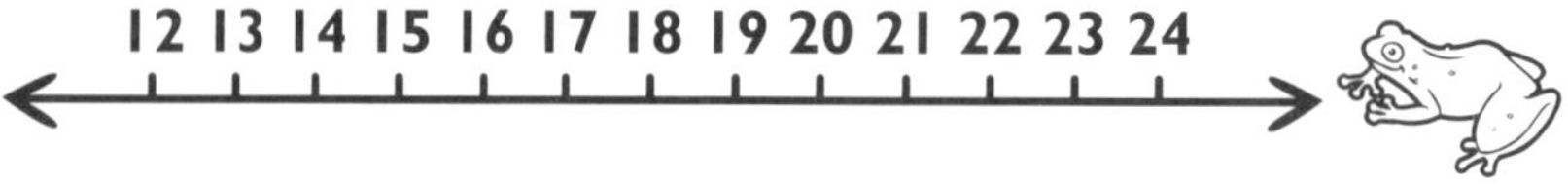

A frog wanted to work out 16 plus 5 by using a number line. It started at 16 and jumped forward 5 numbers. Which number did it land on? How did you work it out?

Encourage students to explain their counting strategies. Invite a student to show the jumps on the number line. Repeat the activity for: 13 + 10, 18 + 3, and 19 + 5.

Grasshopper subtraction

Show a number line from 12 to 24:

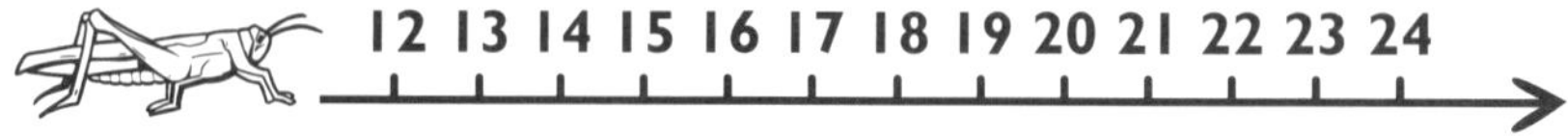

A grasshopper wanted to work out 22 subtract 4 by using a number line. It started at 22 and jumped backwards 4 numbers. Which number did the grasshopper land on? How did you work it out?

Encourage students to explain their counting strategies. Invite a student to show the jumps on the number line. Repeat the activity for: 18 − 5, 20 – 5, 21 − 8.

Frames to 20

Materials

- ten-frames

Briefly show a ten-frame with 10 dots and a ten-frame with 7 dots:

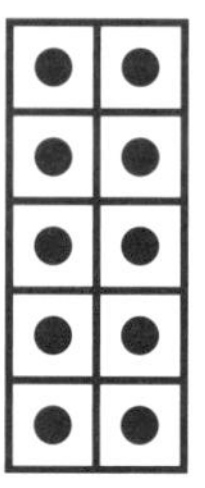

How many dots did you see altogether? How many more dots are needed to make 20? How do you know?

Show this number sentence: **17 + 3 = 20**

Repeat the activity, showing students other ten-frames for numbers 11 to 19.

Combinations for 20

Show these numbers: **3, 13, 8, 2, 7, 6, 14**

Can you see two numbers that add together to make 20? What are they?

Observe whether students can instantly recall number facts for 20.

How did you know which two numbers add together to make 20? (I tried 13 and 6 but that made 19, so I knew it was 1 more – 13 and 7. 3 and 7 make 10 so 13 and 7 make 20.)

Discuss the students' strategies. Repeat the activity often using different numbers.

Memory for 20

Materials

- a set of number cards from 0 to 20 (include two cards for number 10)

Divide the class into three teams. Ask the students to sit in a circle. Tell the students that you are going to play a game with them.

Rules for memory for 20

1. Spread the number cards face down on the floor.
2. In turn, choose a student from each team to turn over two cards.
3. If the two numbers add together to make 20, then the team keeps the cards. If not, the cards are returned face down to the floor.
4. Play continues until all cards are collected. The team with the most cards is declared the winner.

Find a partner

Materials

- a set of number cards from 1 to 5, and 15, 16, 17, 18 and 19

Choose 10 students. Give each student a number card. Ask the students to find a place in the room and hold their cards for all to see.

Can you find the number card that when added to your card makes 20?

When the students have found their partner, ask the remaining students to check the combinations.

Does 16 and 4 make 20? How can we check? (6 and 4 makes 10, and 10 more makes 20. We can count from 16: 17, 18, 19, 20.)

Choose another 10 students and repeat the activity.

Bucket bundles

Materials

- 27 craft sticks
- a bucket

Place a bundle of 10 craft sticks in the bucket. Tell the students that there are 10 craft sticks in the bucket. Place another bundle of 10 craft sticks in the bucket.

How many craft sticks are in the bucket altogether?

One at a time, place 7 more craft sticks in the bucket.

How many sticks are now in the bucket altogether? How many bundles of 10 sticks are in the bucket? How many sticks do I need to put in to make 30 sticks altogether?

Discuss the students' strategies.

Five dice add

Materials

- 5 number dice

Invite five students to stand. Give each student a number dice. Ask these students to roll their dice at the same time.

What is the total of the 5 dice? How did you get your answer? (6 and 4 is 10. Double 2 is 4. 14 and 6 is 20.)

The first student to call the answer remains standing. The other students sit, and new competitors are chosen. Discuss the student's addition strategies. Continue to play until all students have had a turn. Where possible, introduce the idea of combining numbers that add to 5 or 10. For example, if 3 + 6 + 2 + 4 + 1 is thrown, the numbers can be grouped mentally as 6 + 4, 3 + 2 and 1.

Broken calculator

Materials

- a class set of calculators
- pencils
- paper

How could you use a calculator to add 10 and 7 if the 1 button is broken? (8 + 2 + 7; 7 + 3 + 7; 5 + 5 + 5 + 2)

Encourage students to work in pairs to record as many possible combinations as they can. Observe whether students use a pattern to organise their answers. Repeat the activity for other 2-digit numbers. Students can demonstrate their solutions using a calculator

New bag

Elena wants to buy a new school bag that costs $20. Her parents gave her $12 and she has $9 in her money box. Does Elena have enough money to buy the bag? How do you know?

Invite the students to work in pairs and discuss their strategies. Choose students to present their findings. Discuss the strategies used.

Subtraction dots

Materials

- ten-strips

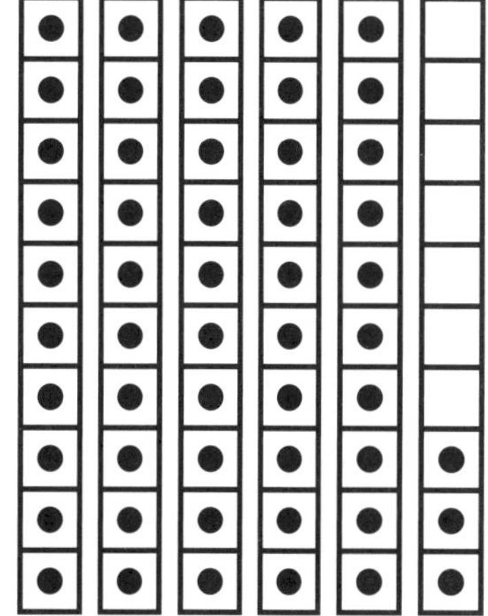

Show five ten-strips with 10 dots and another ten-strip with 3 dots:

How many dots can you see? (53) How many groups of 10? How many ones? How many dots will there be if I take a group of 10 dots away?

Remove one ten-strip.

How many dots are there now? (43) How do I write that number? How did the 2-digit number change?

Year 2

Hidden numbers (1)

Materials

- a bucket
- Unifix or Multilink cubes

Ask the students to sit in a circle. Place a bucket upside down in the middle. Place three towers of 10 cubes on top of the bucket. Tell the students that there are 30 cubes. Ask the students to close their eyes. Place 17 of the 30 cubes under the bucket.

How many blocks are now on top of the bucket? Can you work out how many blocks I have placed under the bucket? How did you work it out? (13 + 17 = 30. 7 more makes 20, then you need 10 more to make 30. I counted from 13 to 30: 14, 15, 16 . . . I counted back from 30 to 13: 30, 29, 28 . . .)

Discuss the students' strategies.

Hundreds chart

Materials

- a hundreds chart

Invite a student to locate number 46 on the hundreds chart:

What is 46 – 1? What is 46 + 1? Where are the answers on the hundreds chart?

Show these numbers: **+10 –10 +1 –1**

Ask the students to choose a 2-digit number from the hundreds chart. Ask them to use the hundreds chart to add and subtract 10 and 1 from their number and record the answers as number sentences, for example. **34 + 10 = 44 34 – 10 = 24**

34 + 1 = 35 34 – 1 = 33.

1	2	3	4	5	6	7	8	9	10
11	12	13	14	15	16	17	18	19	20
21	22	23	24	25	26	27	28	29	30
31	32	33	34	35	36	37	38	39	40
41	42	43	44	45	46	47	48	49	50
51	52	53	54	55	56	57	58	59	60
61	62	63	64	65	66	67	68	69	70
71	72	73	74	75	76	77	78	79	80
81	82	83	84	85	86	87	88	89	90
91	92	93	94	95	96	97	98	99	100

Adding 9

Show the following number sentence: **37 + 10**

How could you work out this problem?

Ask a group of four students to show 30 fingers. Ask another student to show 10 fingers. Ask another student to show 7 fingers to get to the answer.

What if we wanted to add 9 to 37, have we added too much or too little? Could you use 37 + 10 to work out 37 + 9?

If necessary, use a number line to demonstrate adding 10 to 37 then 1 less. Repeat the activity with the following number sentence: **27 + 10.**

Make and check

Materials

- Unifix or Multilink cubes or Base 10 materials
- number lines from 0 to 100

There were 93 students in a Year 2 classroom. Then 25 students left the classroom to attend an athletics carnival. How many students were left in the classroom? Can you work out this problem using your own methods? How could you check your answer?

Encourage the students to use a variety of solution strategies. Develop the method of using the opposite operation to verify the answer, that is, 93 – 25 = 68 as 68 + 25 = 93.

Number combinations (1)

Two numbers add up to 12. Can you work out what the numbers might be? (10 + 2; 6 + 6; and 8 and 4)

Allow students to work in pairs to record the possible solutions.

How do you know you have all possible combinations? Can you use patterns to order your answers?

Discuss the students' responses.

Number combinations (2)

Two numbers add together to make 20. What could the two numbers be? (12 and 8; 14 + 6)

Allow students to work in pairs to record the possible solutions.

How do you know you have all possible combinations? Can you use patterns to order your answers? What do you notice about the numbers?

Discuss the students' responses.

Using doubles

Show these number sentences: **8 + 7 =** ______ **8 + 8 = 16** **7 + 7 = 14**

How can we use 8 + 8 = 16 and 7 + 7 = 14 to complete the number sentence 8 + 7 = ______? (8 + 7 is 7 plus 7 and 1 more. 8 + 7 is the same as double 8 take away 1.)

Encourage students to suggest possible ways of using doubles to solve near-doubles problems. Repeat the activity using other near-double additions, such as **12 + 13   15 + 14   22 + 23.**

Comparing numbers

Materials

- Unifix or Multilink cubes

Chanya and Alex were saving to buy rollerblades. Chanya had saved $28 and Alex had saved $17. How much more money did Chanya have than Alex? How did you work out this problem? (I started at 17 and counted on to find out how many more to make 28. I took 17 away from 28.)

Discuss the different strategies used by students. Invite a student to count 17 cubes and make a tower. Invite another student to count 28 cubes and make a tower. Place the towers side by side to compare. Encourage students to describe how they could use the towers to find the difference between 17 and 28.

Enough money?

Rania wants to buy a phone cover that costs $30. She has $19 in her bag and $12 in her change jar. Does Rania have enough money to buy the phone cover? How do you know?

Invite the students to work in pairs and discuss their strategies. Choose students to present their findings. Discuss the strategies used.

Circle addition

Materials

- Unifix or Multilink cubes

Ask the students to sit in a circle. Choose six students around the circle to each hold a tower of 10 cubes and give one cube to each of the remaining students. In turn, each student places their cubes in the middle as the class makes an oral count of the total number of cubes.

Let us count. 1, 2, 3, 4, 14, 15, 16, 17, 18, 19, 29, 30, 31, 32, 33, 34 . . .

Variation

Start with the total number of cubes on the carpet. In turn, students pick up their cubes from the middle as the class keeps a progressive count backwards.

Bucket addition

Materials

- 36 craft sticks
- a bucket

Place one bundle of 10 craft sticks in the bucket. Tell the students that there are 10 craft sticks in the bucket. One at a time, place another two bundles of 10 craft sticks and then 6 single craft sticks in the bucket.

How many craft sticks are there in the bucket altogether? How did you work it out? (26 plus 10 equals 36. 1 ten plus 2 tens and 6 more is 36. I counted 10, 20, 30, 36.)

Invite students to share the strategies they used to find the solution. Use the craft sticks to model some of the strategies.

Bucket takeaway

Materials

- 20 craft sticks
- a bucket

Show the students a bucket. Count 20 craft sticks into the bucket.

There are 20 craft sticks in the bucket. How many craft sticks will be in the bucket if I take out 14? (Take 10 craft sticks and then take 4 to leave 6. I started with 20 and counted 14 backwards by ones. The answer is 6 because you need 6 more than 14 to make 20.)

Invite students to share the strategies they used to find the solution. Use the craft sticks to model some of the strategies.

Broken calculator

Materials

- a class set of calculators

How could you use a calculator to add 20 and 6 if the [2] *button is broken? (16 + 4 + 6; 19 + 1 + 6; 17 + 3 + 6)*

Encourage students to work in pairs to record as many possible combinations as they can. Observe whether students use a pattern to organise their answers.

Combining to 10

Show these numbers: **12, 7, 3, 8**

Can you see a quick way to add these numbers? (Do 7 + 3 first to make 10, then 12 and 8 makes 20, so 20 + 10 = 30.)

Explain to students that the numbers can be added in any order. Ask the students to check this statement by changing the order in which they add the numbers. Record all possible number sentences.

Subtraction patterns

Materials

- a hundreds chart

Show the students a hundreds chart:

1	2	3	4	5	6	7	8	9	10
11	12	13	14	15	16	17	18	19	20
21	22	23	24	25	26	27	28	29	30
31	32	33	34	35	36	37	38	39	40
41	42	43	44	45	46	47	48	49	50
51	52	53	54	55	56	57	58	59	60
61	62	63	64	65	66	67	68	69	70
71	72	73	74	75	76	77	78	79	80
81	82	83	84	85	86	87	88	89	90
91	92	93	94	95	96	97	98	99	100

Invite a student to point to number 52 on the chart.

Can you tell me a way to subtract 10 from 52 using the chart? What is the answer? What if I subtract 20 from 52? What if I subtract 40 from 52? What do you notice? (To subtract 10, you go up one square.)

Discuss the pattern with the students. Discuss the students' responses.

Guess my number

Materials

- a hundreds chart

Show the students a hundreds chart.

Ask the students to guess the number, giving these clues:

When I add 10 to this number, the answer is 78. When I take away 1 from this number, the answer is 67. When I take away 10 from this number, the answer is 58.

Discuss the students' strategies. Invite students to pose similar questions for others to determine their mystery number.

Addition dots

Materials

- ten-strips

Show one ten-strip with 10 dots and another ten-strip with 4 dots:

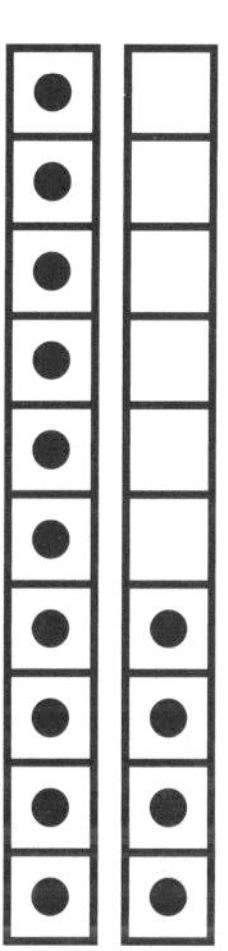

How many dots can you see? How can you count the dots without counting by ones? (1 ten and 4 ones)

Discuss the students' strategies for counting the dots quickly.

How many more dots do I need to make 20 dots? How do you know? If I want to show 23 dots, how many more dots do I need? How did you work it out? (You need 6 to make 20 and 3 more.)

Encourage students to explain their strategies.

Subtraction dots

Materials

- ten-strips

Show seven ten-strips with 10 dots and another ten-strip with 3 dots:

How many dots can you see? (73) How many groups of 10? How many ones? Can you tell me a way to take 10 away from 73 using the dots?

Remove one ten-strip.

How many dots are there now? Which digit has changed? Which digit will change if we took away another 20 dots? Which digit will change if we took away 2 dots?

Discuss the students' strategies.

Hidden numbers (2)

Materials

- ten-strips
- a piece of paper

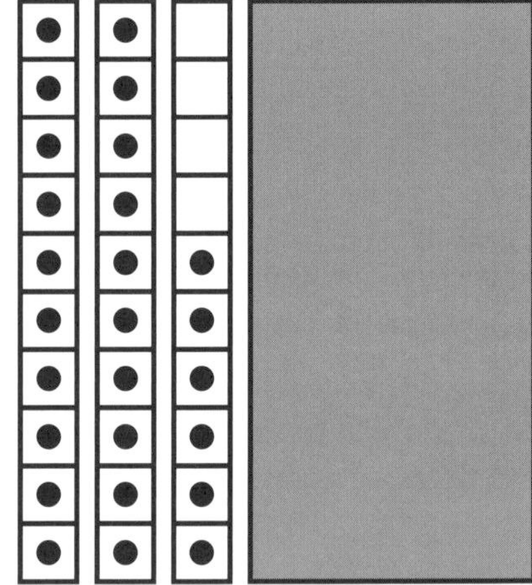

Show two ten-strips with 10 dots and another ten-strip with 6 dots next to a piece of paper:

Which number is shown here? (26. 2 tens and 6 ones)

Tell the students that you have placed some other ten-strips under the paper.

If I have another 20 dots under the paper, how many dots are there altogether? How did you work it out? (I started at 26 and counted 20 more by ones. I added 20 and 20 to make 40 and 6 more.)

Discuss the students' strategies.

Hidden numbers (3)

Materials

- ten-strips
- a piece of paper

Show one ten-strip with 10 dots and another ten-strip with 4 dots next to a sheet of paper:

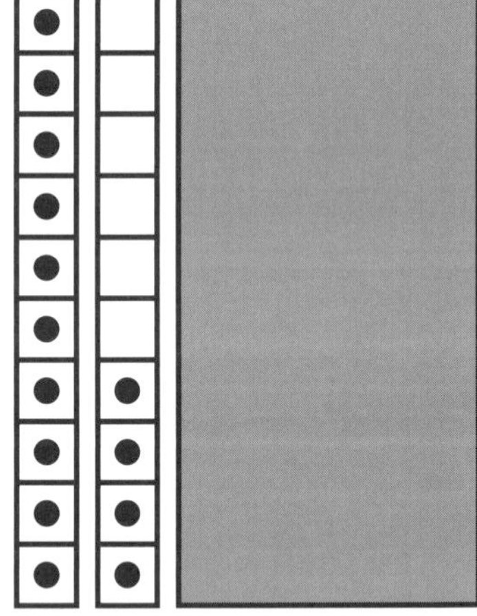

Which number is shown here? (1 ten and 4 ones. 14)

Tell the students that you have placed some other ten-strips under the paper.

If I have 30 dots altogether, how many are hidden? How did you work it out? (7 more to make 20, then 10 more to make 30. I counted from 14 to 30: 15, 16, 17 . . . I counted back from 30: 30, 29, 28 . . .)

Discuss the students' strategies.

Add and check

Show this number sentence: **24 + 10**

How could you work this out?

To model this problem, ask three students to stand and show 24 fingers between them. Ask another student to stand and show 10 fingers. Combine the 20 and 10 to make 30. Add 4 to obtain the answer 34.

What if we wanted to add 11 to 24, have we added too much or too little? How could you use 24 + 10 to work out 24 + 11? (We would need to add 1 more.)

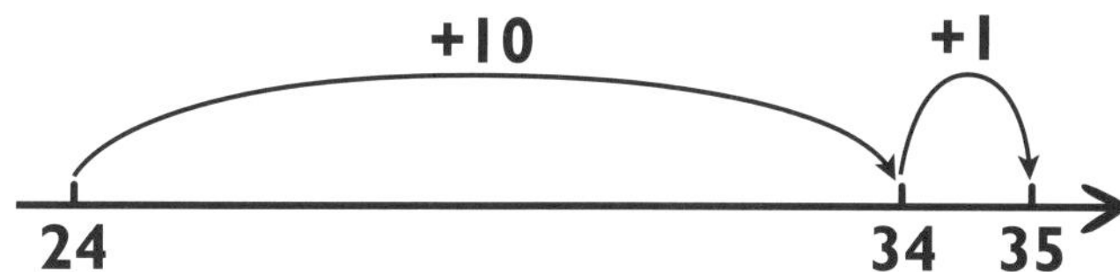

Use a number line to demonstrate adding 10 to 24 and 1 more.

Variation

Show this number sentence: **43 + 11.**

Ask the students to explain how they would work it out.

Take away and check

Show this number sentence: **26 – 10**

How could you work this out?

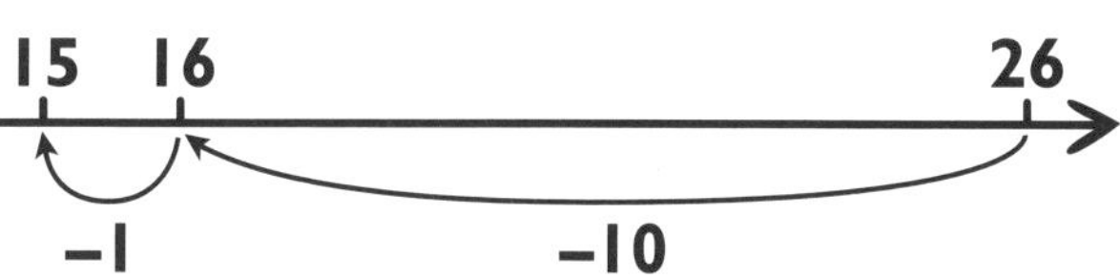

To model this problem, ask three students to stand and show 26 fingers between them. Ask one student to hide their fingers to take away 10 and obtain the answer 16.

What if we wanted to take away 11 from 26, have we hidden too much or too little? How could you use 26 – 10 to work out 26 – 11? (We would need to take away 1 more.)

Use a number line to demonstrate taking 10 from 26 and 1 less.

Variation

Show this number sentence: **46 – 11**. Ask the students to explain how they would work it out.

Addition bridging (1)

Show these numbers: **80, 50, 30, 70, 40, 20, 60, 10**

What do you notice about these numbers? (They all end with 0.)

Tell the students that these are called decade numbers.

What is the first decade number after 19? . . . 37? . . . 62? . . . 75? Which number would you add to 16 to make the next decade number? Which number would you add to 58 to make the next decade number?

Discuss the students' strategies.

Addition bridging (2)

Show this number sentence: **34 + 7**

How could we work out 34 + 7? (We could start at 34 and keep counting by ones, but that would be slow. 34 plus 6 is 40, add 1 more is 41.)

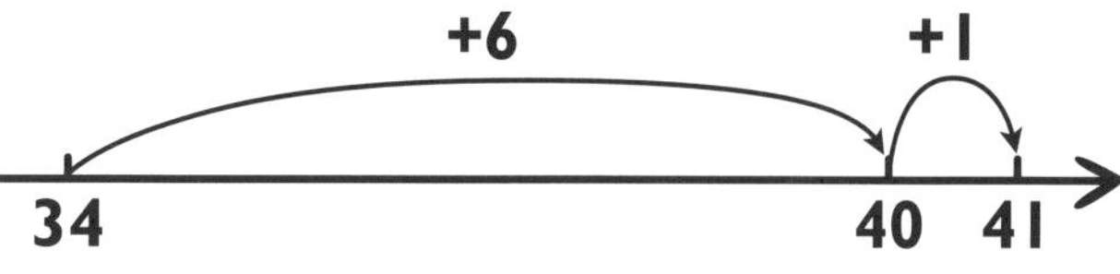

Use a number line to record the 'bridging to tens' strategy.

Repeat this process for other number sentences: **46 + 8** **25 + 7** **37 + 4** **59 + 8.**

Subtraction bridging (1)

Show these numbers: **80, 50, 30, 70, 40, 20, 60, 10**

What is the first decade number before 27? Which number would you take from 27 to make that decade number?

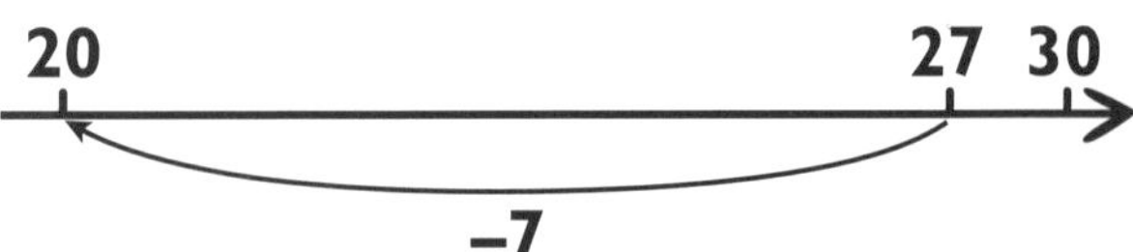

Use a number line to record the 'bridging' strategy.

Repeat this process for other numbers.

Subtraction bridging (2)

Show this number sentence: **25 – 7**

How could we work out 25 – 7? (We could count back by ones from 25. We could take away 5 then take away 2 more.)

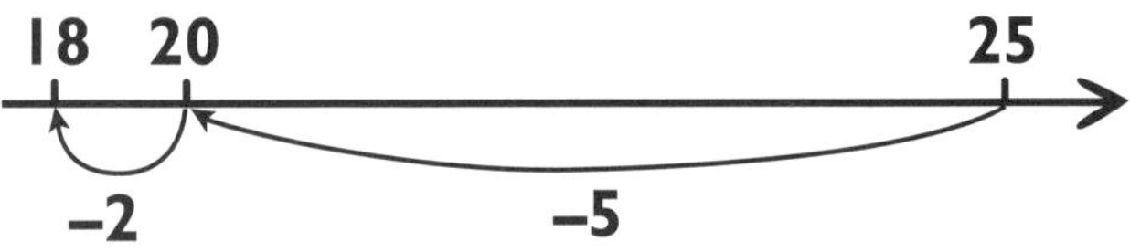

Use a number line to record the 'bridging to tens' strategy.

Repeat this process for other number sentences: **27 – 9** **34 – 6** **18 – 9** **34 – 9.**

Number sentences

Show these numbers: **32, 2, 29, 3, 26, 27**

Can you use any two of these numbers to make a number sentence? How many different addition and subtraction number sentences can you make?

Encourage students to work in pairs to record their work. Check their answers and discuss their strategies.

Variation

Repeat the activity, adding or subtracting three numbers in each number sentence. Mixed addition and subtraction may also be used as students become more proficient.

Three for 100

Materials

- a class set of calculators

Show this number sentence: ______ **+** ______ **+** ______ **= 100**

Can you tell me three numbers that add up to 100?

Record the students' responses. Encourage students to work in pairs to record additional solutions. Calculators may be used.

Number show

Materials

- a set of 2-digit number cards (20, 25, 29, 30, 35, 39, 40, 45, 49, 50, 55 and 59)
- a set of number cards from 0 to 9

Divide the class into two teams. Choose one student from each team.

Rules for number show

1. Give one student the 2-digit cards and the other student the 1-digit cards.
2. Both students turn over the top card, attempt to add the two numbers together and call out the answer.
3. The student who states the correct answer first must explain their strategy. Their team earns a point for each correct answer.
4. Choose pairs of students until the game is over.

Variation

Play again, subtracting the 1-digit number from the 2-digit number.

Addition strategies

Show this diagram with this number sentence in the middle:

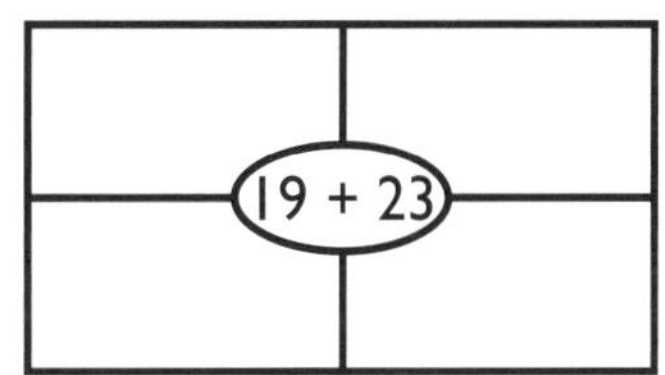

Can you tell me four different ways to add 19 + 23? (20 + 23 is 43, and 43 – 1 is 42. I started at 23 then counted on 19 more. 10 + 20 is 30, 30 + 9 is 42, 23 + 10 is 33 + 9 is 42.)

Discuss the different strategies suggested by students. Record each response in a separate section of the diagram. Repeat the activity often for combinations of other 2-digit numbers.

Subtraction strategies

Show this diagram with a number sentence in the middle:

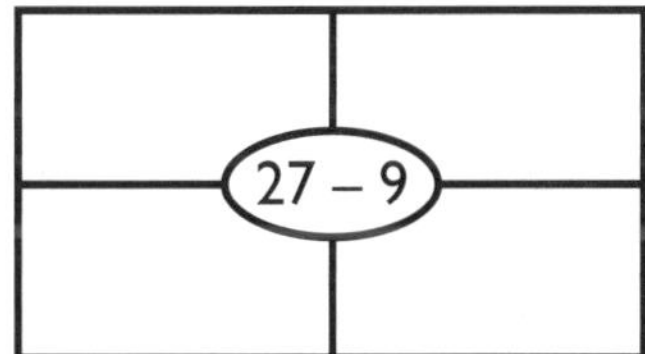

Can you tell me four different ways to solve 27 – 9? (27 take away 7 is 20, then take away 2 more. 27 take away 10 is 17, then put 1 back to get 18. I counted from 9 to 27 to get 18. I counted back 9 from 27 to get 18.)

Discuss the different strategies suggested by students. Record each response in a separate section of the diagram. Repeat the activity often for combinations of other 2-digit numbers.

Multiplication and division

Year 2

Count the tomatoes

Materials

- counters

Bashur, Aidan and Jakob were picking tomatoes. They each had a small bag containing 5 tomatoes. How many tomatoes did they have altogether?

Encourage the students to close their eyes and picture the number of tomatoes.

Can you use the counters to show the class how many tomatoes are there?

Invite three students to replicate the story using counters. Ask the other students to check their answers. Arranging the groups in arrays is not necessary at this stage, although if it arises it can be discussed.

Count the sausage rolls

At a party I saw 3 plates on a table. Each plate had 4 sausage rolls on it. How many sausage rolls were on the table? Can you picture the plates and the sausage rolls?

Encourage the students to close their eyes and picture the groups.

Can you tell me how many sausage rolls you can see? (4 . . . 5, 6, 7, 8 . . . 9, 10, 11, 12; 4, 8, 10 . . . 12 altogether.)

Discuss the students' counting strategies. Show 3 groups of 4 sausage rolls and model them as equal groups.

Making body part groups

Encourage students to find a space of their own in the room.

Can you join with other students to make a total of 10 eyes?

Allow enough time for the activity.

How did you count to check that there are 10 eyes? (We counted 1, 2, 3, 4, 5, 6, 7, 8, 9, 10; 2 + 2 + 2 + 2 + 2 is 10. We counted 2, 4, 6, 8, 10.) Could you check that there are 10 eyes another way? How many people were needed? (We know that 5 twos are 10 so we needed 5 people.)

Discuss the students' responses.

Variation

Repeat this activity for other body parts. Join with others to make a total of 14 arms, 16 shoes, 20 fingers and 40 toes.

Making cubes

Materials

- Multilink cubes

Give each student 4 interlocking cubes.

Can you join with others to make cubes with 12 blocks? How many people are needed to make 12 cubes? How could you count to check that there are 12 cubes? (We counted them one at a time: 1, 2, 3 . . . We counted 1, 2, 3, 4, 5, 6, 7, 8, 9, 10, 11, 12.)

Show the students how to arrange their cubes as an array.

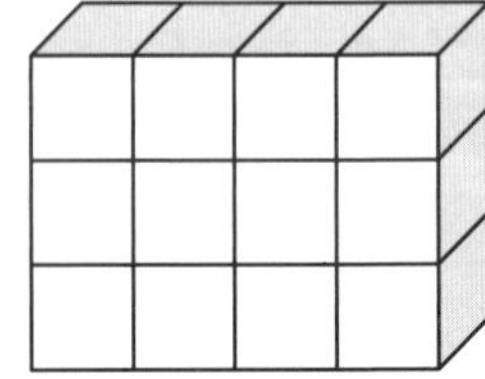

Is there another way to count the cubes? (4 plus 4 plus 4 is 12. We know that 3 groups of 4 is 12.)

Discuss the students' responses.

Variation

Repeat the activity, asking students to join with others to make 16, 20 and 24 cubes.

Kangaroo jumps

Show this number line: 0 ——— 30

Two kangaroos are playing on a number line. The grey kangaroo can jump forward by twos. The red kangaroo can jump forward by fives. If the kangaroos start at 0, will they both land on 30? (If the grey kangaroo jumps by 3, it will land on 30. If the red kangaroo jumps by 5, it will land on 30 too.)

How many jumps will each kangaroo take? How did you work it out?

Discuss the students' counting strategies.

Extension

Which other numbers could the kangaroos jump forwards by to land on 30?

Encourage students to work in pairs to explore a variety of solutions. Students could use a calculator to investigate these solutions.

Skip counting

Materials

- counters

There are 8 soccer shirts on the clothesline and 2 pegs have been used to hang each shirt. How many pegs have been used altogether? How did you work it out? (8 groups of 2 is 16.)

Can you use skip counting to work it out? ($2 + 2 + 2 + 2 + 2 + 2 + 2 + 2 = 16$) Can you suggest a number sentence that describes the problem? ($8 \times 2 = 16$)

Invite students to use counters to represent their solutions.

How many chairs?

Materials

- counters
- craft sticks
- Unifix or Multilink cubes

A school has 4 classrooms. Each classroom has 21 chairs. How many chairs are there altogether? That is 4 groups of 21. How did you work it out?

Provide manipulative materials for students to demonstrate their thinking.

Can you suggest a multiplication number sentence for the problem? (It is 4×20 and add 4 more.) How can you use addition to work it out?

Emphasise the use of addition to solve multiplication problems.

Lettuce arrays

Materials

- 16 counters

Mr Greenthumb has 16 lettuce seedlings. He wants to plant them in equal rows. How many ways can Mr Greenthumb plant his seedlings?

Discuss the different arrangements. Invite the students to model their arrays with counters.

How can you count the number of counters in this array? (Count by ones: 1, 2, 3, 4, 5, 6, 7, 8 . . .) Can you count them another way? (Count by fours: 4, 8, 12, 16 . . .)

Discussing arrays

Materials

- 15 counters

Show 15 counters in an array of 3 rows of 5:

How many counters are there altogether? How did you work it out? Can you work it out another way?

Discuss the idea of counting equal rows in an array to obtain the answer. Focus on division.

How many groups of 3 counters can be made from these? How did you work it out? How many groups of 5 counters can be made from these? How did you work it out?

Discuss the idea of sharing objects into equal rows in an array.

Quick arrays

Materials

- counters

Arrange the counters as an array of 4 rows of 5. Show the array briefly then cover it:

What did you see? (I saw 5 fours. I saw 4 rows of 5 counters. I saw 4 fives.) How many counters were there altogether? How did you work it out? (5 and 5 and 5 and 5 make 20.)

Encourage students to describe the array in their own way. Show the array again. Model the strategies that students used to determine the total number of counters. Repeat the activity using different arrays.

Counter arrays

Materials

- 12 counters for each pair of students
- pencils
- paper

Invite the students to work in pairs for this activity.

How many different arrays can you make using 24 counters?

Give each pair of students a piece of paper to record their solutions. Discuss the different arrangements. Invite students to model the arrays using counters.

Can you describe your array?

Compare a 6 × 4 array and a 4 × 6 array. Focus on the commutative property.

What do you notice about these solutions?

Discuss the idea that, when multiplying, the order of the operation does not affect the answer.

Estimating area

Materials

- different-sized rectangles
- counters

Show a rectangle with a counter on it:

Can you estimate the number of counters needed to cover this rectangle?

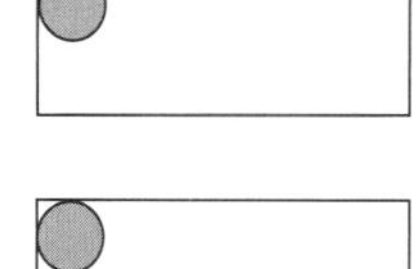

Discuss the students' strategies. Complete a row or column of counters and encourage the students to review their estimation.

Repeat the discussion. Completely cover the rectangle to find the actual number of counters required. Repeat the activity for different-sized rectangles.

Multiple frames

Materials

- ten-frames

Briefly show four ten-frames with 5 dots each:

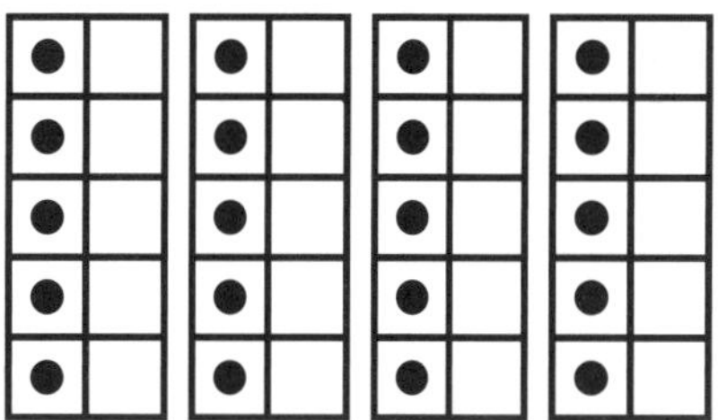

Can you describe what you saw? (I saw 5 and 5 and 5 and 5. Four groups of 5 counters.) How many dots are there altogether? How did you work it out? (2 tens are 20. 5 + 5 is 10 and double ten is 20.)

Repeat this activity with a different numbers of dots on the ten-frames.

Showing triangles (2)

Show a row of 10 triangles. Tell the students that they are going to use the triangles to help them to skip count by threes.

How many sides can you see on each triangles? How many sides on all the triangles in total?

Record the progressive total below each triangle.

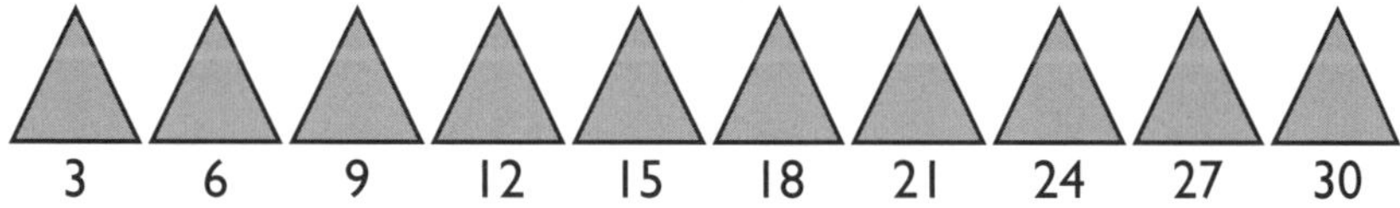

How many triangles make a total of 21 sides? How many sides on 5 triangles? How many triangles would be needed to make 39 sides? What is the pattern that you notice here?

Encourage students to explain how their answers were obtained.

Describing arrays

Materials

- 15 counters

Show 15 counters in an array of 3 rows of 5:

Can you describe this array? (3 rows of 5. 5 columns of 3.) Can you suggest some multiplication and division number sentences for this array? (15 divided by 3 is 5 and 15 divided by 5 is 3.)

Record students' suggestions using symbols, such as $5 \times 3 = 15$, $3 \times 5 = 15$, $15 \div 3 = 5$, $15 \div 5 = 3$.

Sharing equally

Materials

- 5 bundles of 10 craft sticks
- 8 craft sticks

Ask the students to sit in a circle. Choose two students to stand. Place 5 bundles of 10 craft sticks and 8 single craft sticks in the middle of the circle.

How could we share these craft sticks equally between the two students? (We need to break up a bundle.) How many will each student get? How did you work it out? (Two bundles each and one left over, split between them, then split the single ones into four each.)

Encourage students to suggest ways to share the craft sticks. Model some of the suggestions using the craft sticks. Repeat the activity for 48 craft sticks and three students.

Sharing snakes

Materials

- counters

There are 24 lolly snakes in a packet. If the snakes are shared equally among 8 children, how many will each child get?

Encourage students to use drawings, counters and numerals to solve the problem.

How did you work it out?

Invite the students to model their strategies using counters.

Repeated subtraction

Show this number line: 0 ... 24

A frog was playing on a number line.
The frog only likes to jump backwards. If the frog started at number 24, will it land on 0 if it jumps by twos? How did you work it out? How many jumps will it take?

Discuss the students' strategies.

Which other numbers could the frog jump backwards by to land on 0? (If it jumped backwards by 4, it would land on 0. And 24 – 8 – 8 – 8 = 0.) How many jumps will it take?

Encourage students to explore a variety of strategies.

How many biscuits?

Materials

- 20 counters

Place 20 counters in a random arrangement.

Mr O'Brien baked 20 biscuits to share among his children. Each child gets 4 biscuits. How many children will get biscuits? How did you work it out? (I worked out how many fours in 20. I imagined taking 4 away until there were none left.)

Invite some students to each take 4 counters from the collection. Continue this until there are no counters left. Encourage students to suggest a number sentence to represent this problem (e.g. $20 - 4 - 4 - 4 - 4 - 4 = 0$ or $20 \div 4 = 5$). Highlight the connection between multiplication and division by asking students to suggest a multiplication sentence to represent the problem (e.g. $4 \times 5 = 20$ or $5 \times 4 = 20$).

Counting back

Materials

- a class set of calculators

Show this number: **60**

How many times can you take 5 away from this number until you reach 0? How did you work it out?

Discuss the students' strategies. Use a calculator to demonstrate how to repeatedly subtract 5 from 60. (Press [6] [0] [–] [5] [=] [=] [=] . . .) Encourage students to work in pairs to practise this procedure for other multiples of 5.

Variation

Extend the activity to multiples of 2 and 10.

FRACTIONS AND DECIMALS

Year 1

Sharing counters

Materials

- 12 counters

Show the students 12 counters and ask a pair of students to share the counters between them.

How many counters will you get if you share them equally? How do you know they are equal shares? How many counters did you get?

Repeat the activity for groups of three, four, five and six students so that they experience the concept of equal and unequal grouping.

How many ways can you make equal groups? (6 groups of 2, 2 groups of 6, 3 groups of 4, 4 groups of 3). What do you notice about these groups?

Discuss the students' responses and focus on commutative property.

Sharing craft sticks

Materials

- a bundle of 10 craft sticks
- 8 single craft sticks

Show the students the craft sticks and ask a pair of students to share the craft sticks between them.

How many craft sticks will you get if you share them equally? (4 single craft sticks and some from the bundle) How do you know they are equal shares? How many craft sticks did you get? (4 singles and 5 from the bundle, so that is 9)

Repeat the activity for 15 craft sticks and three students.

Labelling halves

Materials

- paper strips
- paper circles
- paper squares

Give the students a selection of paper strips, circles and squares.

Can you fold your paper into halves? What can you tell me about them? (There are two halves. They are the same size.)

Discuss the students' strategies. Ask them to label their halves using words and symbols, and make a class display.

Halving groups

Materials

- counters

Show this arrangement of counters:

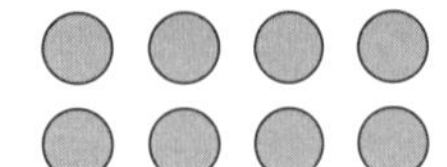

If I had half this number of dots, how many dots would I have? (Half means two equal groups. Half of 8 is 4.)

Invite students to demonstrate their solutions using counters. Ask the students to close their eyes.

Can you imagine what half of a group of 10 dots would look like?

Discuss the students' responses.

Describing parts

Materials

- counters

Show this arrangement of counters:

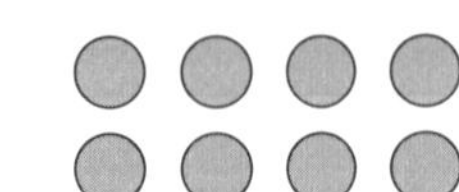

Separate 3 counters from the rest of the collection.

Is this group more than one-half or less than one-half? How do you know?

Discuss the students' responses. Repeat the activity for the group of 5 counters.

Variation

Repeat the activity for 10 and 12 counters.

Shape halves

Show these shapes:

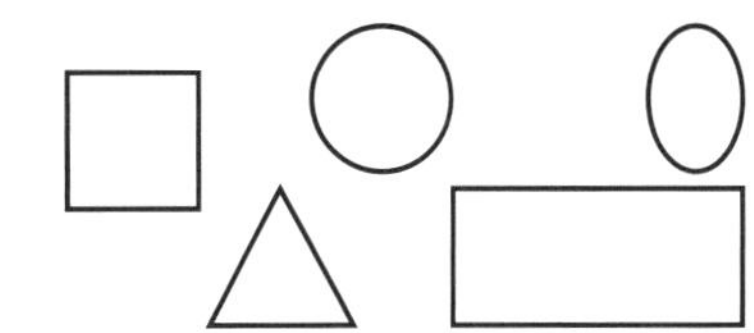

How could you divide these shapes in half?

Choose different students to show their solutions.

Do the shapes show equal parts? How do you know? Can you divide the shapes in half another way?

Discuss the students' responses.

Year 2

Shape quarters

Show these shapes:

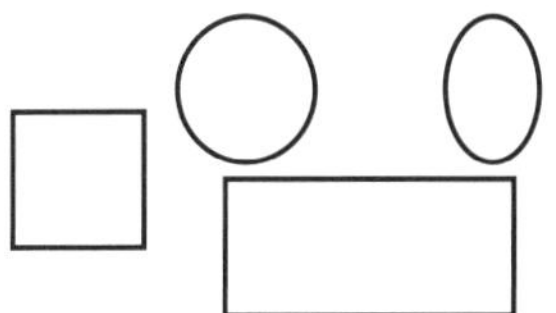

How could you divide these shapes in quarters?

Choose different students to show their solutions.

Do the shapes show equal parts? How do you know? Can you divide the shapes in quarters another way?

Discuss the students' responses.

Labelling quarters

Materials

- paper strips
- paper circles
- paper squares

Hand out a selection of paper strips, circles or squares to the students.

How can you fold your shape into quarters? (You fold the strip/shape in half and then you fold it again.)

Allow students time to complete the activity. Discuss their strategies. Ask the students to label their shapes using words and symbols. Make a show of students' shapes by gluing them to coloured paper.

Choc chip halves

Show a picture of one-half of a choc chip cookie with 8 choc chips. Tell the students that the missing piece also contained 8 choc chips.

How many pieces has the cookie been divided into? (There are two pieces.) What is each piece called? (They are halves.) How many choc chips are there in each half? How many choc chips would there be in a whole cookie?

Discuss the students' responses.

Choc chip quarters

Show a picture of a choc chip cookie divided into quarters with 3 choc chips in each quarter.

How many pieces has the cookie been divided into? What is each piece called? How many choc chips are there altogether? How many choc chips are there in each quarter?

Discuss the students' responses.

Missing quarters

Show a picture of one-quarter of a choc chip cookie with 6 choc chips. Tell the students that each of the missing pieces contained the same number of choc chips.

How many pieces has the cookie been divided into? What is each piece called? How many choc chips are there in each quarter? (There are 6 pieces in each quarter.) How many choc chips would there be in a whole cookie? How do you know? ($6 \times 4 = 24$)

Discuss the students' responses.

In the hand

Materials

- Multilink or Unifix cubes
- a bag

Join 6 cubes together in a row and hide them in a bag. Show the students another row of 6 cubes. Tell the students that half of the cubes are hidden in the bag.

If the other half of the cubes are in my hand, how many cubes are there altogether?

Discuss the students' strategies.

If I gave you one-quarter of the cubes, how many would that be?

Variation

Repeat the activity for 16, 20 and 24 cubes.

Comparing cars

Show a group or picture of 8 toy cars with 2 red cars:

What fraction of the cars is red? (2 out 8 are red. That is the same as $\frac{1}{4}$.)

Circle the 2 cars and show the answer $\frac{1}{4}$.

How many other groups of 2 cars can you see? What are these fractions called? Are they the same as the first quarter? How do you know?

Discuss the students' responses. Develop the idea of the quarters being identical.

Jellybean halves

Materials

- jellybeans

Show a group of 12 jellybeans. Choose two students and share the jellybeans among them, one at a time, until there are none left.

Did each student get an equal amount of jellybeans? How many jellybeans did each student receive? What fraction of the jellybeans did each student receive? How could you check to see whether you were correct?

Discuss the students' responses. Repeat the activity for different-sized groups.

Jellybean quarters

Materials

- jellybeans

Show a group of 16 jellybeans. Choose 4 students from the class and share the jellybeans among them, one at a time, until there are none left.

Did each student get an equal amount of jellybeans? How many jellybeans did each student receive? What fraction of the jellybeans did each student receive? How could you check to see whether you were correct?

Discuss the students' responses. Repeat the activity for different-sized groups.

MONEY AND FINANCIAL MATHEMATICS

Year 1

Australian coins

Look at a set of Australian coins.

Invite students to order a set of coins from the smallest to the biggest in size. Then ask the students to order the coins from the smallest to the largest in value.

What do you notice about their size and shape? Does the size match their value?

Discuss the student's responses.

Australian coin features

Invite students to look closely at a set of Australian coins.

What special features do they have? (Numbers, pictures, colour, patterns, shape, material. . .)

Ask the students to work in groups of six. Each student can look at one coin and describe the features of that coin to their group.

Ten-cent count

Materials

- 10c coins
- a bucket

Ask the students to sit in a circle. Place a collection of 10c coins and a bucket in the middle. Invite the students, one at a time, to place a coin in the bucket.

Can you count the money as each coin is placed in the bucket? (10, 20, 30, 40 . . . 90, 100) What do we call 100c? How many 10c coins make $1?

Discuss the students' responses.

Five-cent count

Materials

- 5c coins
- a bucket

Ask the students to sit in a circle. Place a collection of 5c coins and a bucket in the middle. Invite the students, one at a time, to place a coin in the bucket.

Count the money as each coin is placed in the bucket. (10, 20, 30, 40 . . . 90, 100) What do we call 100c? How many 5c coins make $1?

Discuss the students' responses.

Year 2

Making $1

Materials

- coins

Sabine had two 50c coins in her pocket, Heidi had three 20c coins and four 10c coins and Tran had $1 coin. Who had the most money?

Allow students time to use coin collections or discuss it with a partner.

Are there any other ways we can make $1?

Making $2

Materials

- coins

I have $2 in my purse. What coins might I have? (Four 50c coins . . .)

Discuss the students' responses. Allow students to work in pairs to record as many possible combinations as they can. Some students may benefit from using coins to find combinations.

Making $5

Materials

- coins

I have a total of $5 in coins in my wallet. What coins might I have? ($2 + $2 + $1; $1 + $1 + $1 + $1 + 50c + 50c; $2 + $1 + $1 + 20c + 20c + 20c + 20c + 20c. . .)

Discuss the students' responses. Allow students to work in pairs to record as many possible combinations as they can. Some students may benefit from using coins to find combinations.

Price tags

Materials

- coins

What coins can I use to make these amounts?

How can I make these amounts using the ***least*** *number of coins?*

Discuss the students' responses. Some students may benefit from using coins to find combinations.

PATTERNS AND ALGEBRA

Year F

Repeating patterns

Materials

- counters
- Multilink or Unifix cubes

Show a row of counters of two alternating colours:

What do you notice about these counters? (It repeats.) Which colour will be next in this pattern? (It goes red, blue, red, blue . . .) How do you know?

Discuss the students' responses.

How could we continue this pattern?

Invite students to continue the pattern.

Craft stick patterns

Materials

- craft sticks
- paper
- coloured pencils

Show this craft stick pattern:

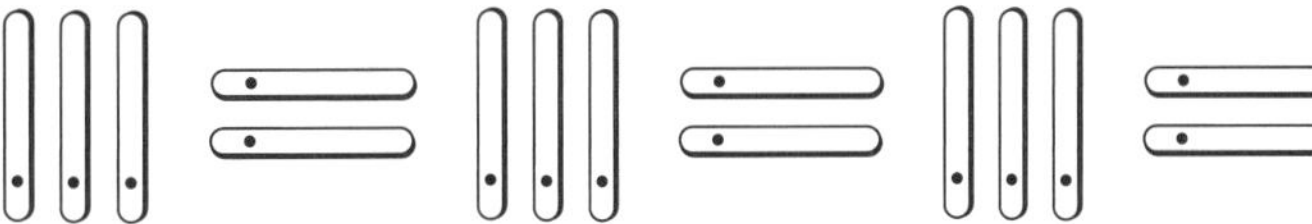

Can you make a repeating pattern using craft sticks? Can you show your pattern on paper?

When students have completed the activity, share and discuss the patterns they have made.

Two-element patterns

Materials

- counters
- Unifix cubes

Show this arrangement of counters, using two alternating colours:

Which colour will be next in this pattern? How could you describe the pattern? (There are 2 yellow, 2 red. It goes by twos.)

Discuss the students' responses. Continue the pattern further. Develop the idea of a '2 pattern', which repeats after every 2 counters or cubes.

Which colour will be the 12th counter in the pattern?

Invite students to continue the pattern.

Three-element patterns

Materials

- counters
- Unifix cubes

Show this arrangement of counters, using three alternating colours:

Which colour will be next in this pattern? How could you describe the pattern? (There is 1 yellow, 1 red, 1 blue. Then it repeats. It is a pattern of threes.)

Discuss the students' responses. Continue the pattern further. Develop the idea of a '3 pattern', which repeats after every 3 counters.

Which colour will be the 12th counter in the pattern?

Invite students to continue the pattern.

Make it three

Materials

- counters

How many different '3 patterns' can we make using counters?

Invite a selection of students to demonstrate their patterns.

How would you describe these patterns? (My '3 pattern' has red, red, blue. . .) Which parts repeat? How many counters are there in each pattern?

Discuss the students' responses.

Variation

Encourage students to create more patterns using manipulative materials such as buttons, coloured craft sticks, coloured blocks, and shapes.

Clap and click patterns

Encourage the students to create a slow rhythmic pattern by clapping their hands and then clicking their fingers. On the first 'clap', ask the students to say the number 1. Invite the students to continue to count as they clap and click.

One (clap), two (click), three (clap), four (click), five (clap), six (click) . . .

Repeat the activity. Choose a starting number in the range from 1 to 20. Stop the count and ask the students to predict the next number.

Variation

Repeat the activity, counting backwards from a number in the range from 10 to 20.
Use different body parts to make claps and taps.

Sound patterns

Materials

- pattern blocks

Show this pattern:

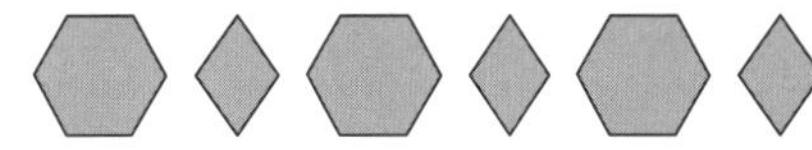

Can you clap and slap to this pattern?
Clap–slap–clap–slap–clap–slap–clap–slap . . .

Invite a student to make another repeating pattern with the pattern blocks. Ask the students to make the new pattern into a sound pattern.

Make a pattern

Invite the students to work in pairs. Tell them that they are going to make their own patterns by clapping, tapping or clicking.

Can you make a '3 pattern'? (My '3 pattern' is clap hands, clap hands and stamp feet. . .)

Have pairs of students demonstrate their '3 patterns' to the rest of the class.

Missing numerals

Show this sequence of numbers: **24, 23, 22,** ______, ______, ______, **18, 17, 16,** ______, ______, ______, **12, 11, 10,** ______, ______, ______, **6, 5, 4,** ______, ______, ______

Can you tell me which numbers are missing from this pattern?

Discuss the students' responses. Count backwards orally with the students.

How many numbers are missing each time?

Share and discuss the students' thinking.

Variation

Introduce other missing number sequences. Include numbers from 1 to 30.

Jumbled patterns (1)

Materials

- pattern blocks

Show this pattern:

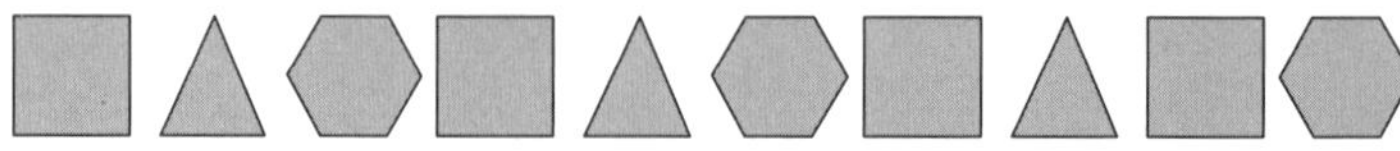

What do you notice about this repeating pattern? (There is a mistake at the end. It should be square, triangle, hexagon.) How could we fix the mistake?

Allow students time to consider the pattern. Discuss the students' thinking. Invite a volunteer to rearrange the pattern so that it is correct.

Other jumbles

Materials

- pattern blocks

Show this pattern:

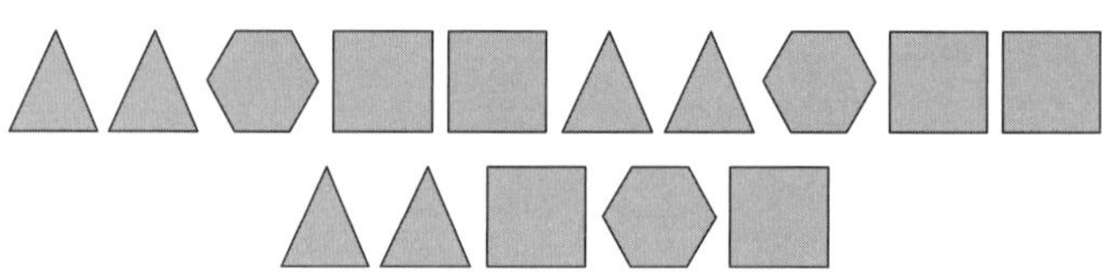

What do you notice about this repeating pattern? (There is a mistake at the end. You need to move the hexagon.) How could we fix up the mistake?

Allow students time to consider the pattern. Discuss the students' thinking. Invite a volunteer to rearrange the pattern so that it is correct. Repeat the activity using other jumbled patterns.

Missing patterns

Materials

- pattern blocks

Show these two arrangements of blocks:

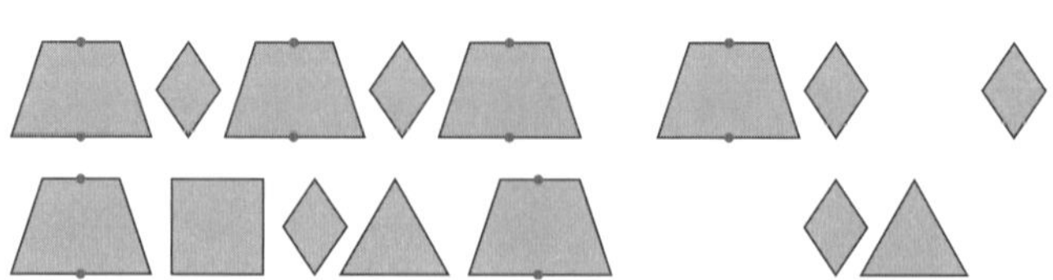

Which shapes are missing from the repeating patterns? How do you know?

Discuss the students' responses. Encourage the students to make some missing patterns for a friend to complete.

Year 1

Missing numbers

Show this sequence of numbers: **28, 27, 26, 25,** ______, ______, ______, ______**, 20, 19, 18, 17,** ______, ______, ______, ______**, 12, 11, 10, 9,** ______, ______, ______, ______, **4, 3, 2, 1**

Can you tell me which numbers are missing from this pattern?

Discuss the students' responses. Count backwards orally with the students.

How many numbers are missing in each part of this pattern? How do you know?

Share and discuss the students' thinking.

Variation

Revise missing 2- and 3-element patterns using numbers to 30.

Jumbled patterns (2)

Materials

- pattern blocks

Show this pattern:

What do you notice about this repeating pattern? (There is a mistake at the end. It should be triangle, 2 squares and then a hexagon.) How could we fix up what is wrong with it?

Allow students time to consider the pattern. Discuss the students' thinking. Invite a volunteer to rearrange the pattern so that it is correct.

Calendar patterns

Materials

- a calendar for the current month

Show the calendar to the students. Encourage students to say the days of the week as you point to them on the calendar. Count the numbers on the calendar show with the students.

Can you see any repeating patterns on the calendar? (The Wednesdays are 2, 9, 16, 23 and 30. . . You count the days of the week and then you start them again.)

Discuss the students' responses.

What day of the week will it be on the 1st, 8th, 15th, 22nd and 29th? How many Thursdays will there be in this month? How many Saturdays and Sundays will there be? How many school days will there be?

Use the calendar to count the days.

Other jumbles

Materials

- pattern blocks

Show this pattern:

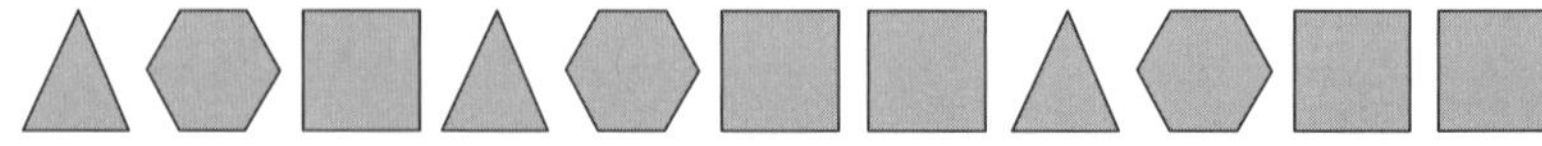

What do you notice about this pattern? (There is a mistake at the beginning, it needs an extra square.) How could we fix it?

Discuss the students' responses and rearrange the pattern so it is correct. Repeat the activity using other jumbled patterns.

Counting colours

Materials

- 3 red counters
- 3 blue counters
- 3 green counters

Show a repeating pattern using coloured counters: red–blue–green–red–blue–green–red–blue–green.

Can you count the number of counters in the pattern?

Record the counting numbers above each of the counters.

Which numbers will be above the next 3 green counters? How did you work it out? What colour will be under number 16? How do you know?

Discuss the students' strategies.

Dot patterns

Materials

- 20 counters for each pair of students

Show these dot patterns:

Give 20 counters to each pair of students.

Can you use your counters to make the next shape in the pattern? How would you describe your shape?

Allow students time to complete their patterns.

How many counters will be in the sixth shape? How did you work it out? Can you predict the number of dots in the seventh shape?

Discuss the students' responses.

Hundreds chart patterns

Materials

- a hundreds chart
- counters

Show the hundreds chart and use counters to cover the first 5 numbers in a number pattern, for example, 2, 4, 6, 8, 10:

Can you describe the number pattern? What is the next number in the pattern? How did you work it out? What will be the 15th number in the pattern?

As a class, count forwards and backwards by the numbers in the pattern. Repeat the activity for other patterns. Students can determine successive numbers by counting or by generalising from the shape of the pattern.

Number card patterns

Materials

- number cards from 1 to 30

Place the number cards face down in order from 1 to 30. Turn over the first four numbers in a number pattern, for example, 2, 4, 6, 8.

Can you describe the number pattern? What is the next number in the pattern? How do you know? What will be the ninth number in the pattern?

As a class, count forwards and backwards by the numbers in the pattern. Repeat the activity for other number patterns.

Counting circle

Ask the students to stand in a circle. Show these numbers: **5, 10, 15, 20, 25, 30**

What do you notice about these numbers? (They are the numbers you say when you count by fives. They all end in 0 or 5.)

Ask the students to start at 1 and count by ones around the circle. Each time a student says a multiple of 5, that student sits down. Continue to count around the circle until one student is left standing. Repeat the activity for other number patterns.

Hopping Hare

Show this number line:

Tell the students that a hare was playing on a number line. It started at 5 and kept jumping forwards by twos.

What numbers will the hare land on?

Invite a student to record the jumps on the number line.

What do you notice about these numbers? (It is jumping on every second number.) Will it land on number 27? How do you know? (It only lands on odd numbers.)

Discuss the students' observations.

Missing number cards

Materials

- number cards for 10, 20, 30, 40, 50, 60 and 70

Place the number cards face down in order from 10 to 70. Tell the students that the numbers on these cards form a number pattern. Invite students, one at a time, to choose a card to turn over. Pause after a few have been turned over.

Do you know the number pattern? Can you work out which numbers are face down? How did you know which numbers they are?

Encourage students to describe the number pattern in their own words.

Year 2

Clapping patterns

Materials

- counters
- Unifix or Multilink cubes

Ask the students to copy a pattern that you clap and stamp. For example, clap-clap-clap, stamp-stamp-stamp . . . . Repeat the activity.

Can anyone show that pattern using counters?

Encourage the students to represent the pattern with different coloured cubes or counters. Repeat the activity and extend the pattern further.

Variation

Encourage the students to model a variety of different clapping, stamping or finger clicking patterns.

Interpreting patterns

Materials

- counters
- cubes

Use coloured counters to make a pattern. For example, red-red, blue-blue, green-green, red-red, blue-blue, green-green. . .

How could you show this pattern using body movements?

Encourage the students to work in pairs to represent the pattern by clapping, stamping or finger clicking. Repeat the activity for a variety of different patterns.

Hundreds chart skips

Materials

- a hundreds chart
- counters

Show a hundreds chart. Ask the students to skip count by fives as you place counters on the numbers from 5 to 100. Count forwards and backwards.

What do you notice about these numbers? (They all end in 0 or 5. They make two columns.)

What happens when you start at 12 and skip count by fives? Why do you think this happens?

Discuss the students' responses. Repeat the activity for counting by tens.

What do you notice about these numbers? What happens when you start at 28 and skip count by tens? Why do you think this happens?

Discuss the students' responses.

Hundreds chart patterns

Materials

- a hundreds chart
- counters

Show the hundreds chart and use counters to cover the first 5 numbers in a number pattern (e.g. 3, 6, 9, 12, 15):

Can you describe the number pattern? What is the next number in the pattern? How did you work it out? What will the 12th number in the pattern be?

As a class, count forwards and backwards by the numbers in the pattern. Repeat the activity for other patterns. Students can determine successive numbers by counting or by generalising from the shape of the pattern.

1	2	●	4	5	●	7	8	●	10
11	●	13	14	●	16	17	18	19	20
21	22	23	24	25	26	27	28	29	30
31	32	33	34	35	36	37	38	39	40
41	42	43	44	45	46	47	48	49	50
51	52	53	54	55	56	57	58	59	60
61	62	63	64	65	66	67	68	69	70
71	72	73	74	75	76	77	78	79	80
81	82	83	84	85	86	87	88	89	90
91	92	93	94	95	96	97	98	99	100

Forwards patterns

Materials

- a class set of calculators
- pencils
- paper

Do you know how to count on using the calculator?

Demonstrate the use of the constant function on a calculator. For example, enter [5] [+] [5] [=] [=] and continue to press [=] so the counting pattern for fives appears. Encourage students to read the numbers shown on the calculator. Ask the students to use the calculator to count on by twos, threes or tens. Encourage students to record these number patterns.

Backward patterns

Materials

- a class set of calculators
- pencils
- paper

Do you know how to count backwards using the calculator?

Demonstrate the use of the constant function on a calculator. For example, enter [1] [0] [0] [−] [5] [=] [=] and continue to press [=] so the decreasing number pattern for fives appears. Encourage students to read the numbers shown on the calculator. Ask the students to start at 100 and investigate counting backwards by twos, threes or tens. Encourage students to record these number patterns.

Dot patterns

Materials

- 20 counters for each student

Show these dot patterns:

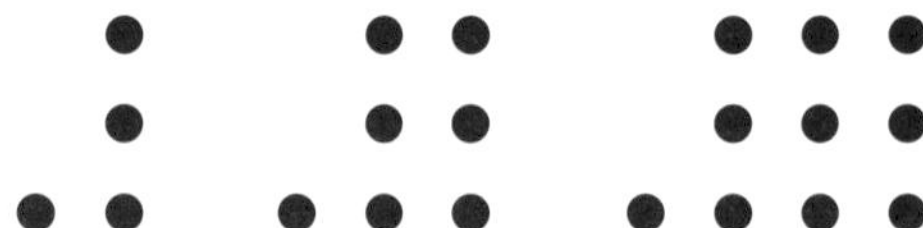

Give 20 counters to each student.

Can you use your counters to make the next arrangement in the pattern? How would you describe your arrangement?

Allow students time to complete their patterns.

How many dots in the fifth arrangement? How did you work it out? Can you predict the number of dots in the sixth arrangement?

Discuss the students' responses.

Missing numbers

Show these numbers: **2, 4, 6, 8, 10, 14, 18, 20**

Tell the students that two numbers are missing from the pattern.

Which numbers are missing? How did you work out which numbers are missing?

Encourage students to describe the pattern.

What will the 15th number in the pattern be? How did you work it out?

Discuss the students' responses. Repeat the activity with a variety of numbers.

Mixed-up numbers

Show these numbers: **13, 5, 7, 9, 3, 15, 17**

Tell the students that these numbers are part of a mixed-up number pattern, but one number is missing.

Which number is missing? How did you work out which number is missing?

Encourage students to determine the first number in the pattern and describe the pattern.

What will the 10th number in the pattern be? How did you work it out? Will 22 be a number in the pattern? How do you know?

Discuss the students' responses. Repeat the activity with a variety of numbers.

Create a pattern

Invite a student to create a number pattern starting at 2 and containing 5 elements. Write the first 5 numbers in the student's pattern, for example, 2, 5, 8, 11, 14.

Can you describe this pattern? What are the next 3 numbers in the pattern? How did you work it out?

Repeat the activity. Choose other students and start the patterns with a variety of numbers.

Decreasing patterns

Show this number line:

0 36

If I started at 36 and kept subtracting 6, what number would I land on? Would I land on 0?

Encourage students to record the numbers they would land on as a number pattern (i.e. 36, 30, 24, 18, 12, 6, 0).

Which other numbers could I repeatedly subtract to land on 0?

Discuss the students' responses.

Balancing addition

Show an equal arm balance with this number sentence on one side:

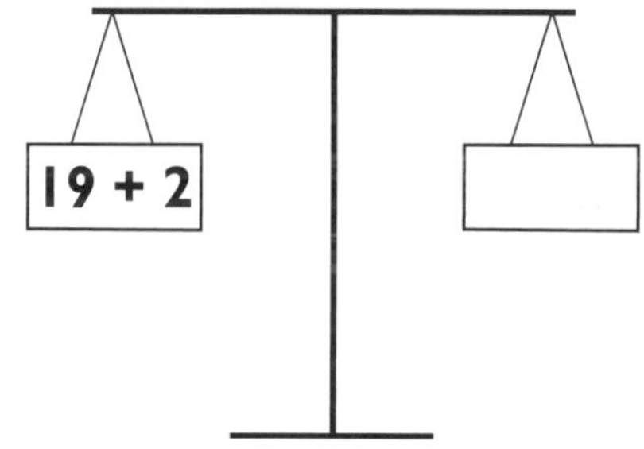

Which other addition facts will make the balance level?

Record the number facts suggested by students on the other side of the balance. Emphasise the idea of equality, that is, that 19 + 2 'is the same as' 17 + 4. Repeat the activity for: 18 + 2 and 15+4.

Balancing subtraction

Show an equal arm balance with this number sentence on one side:

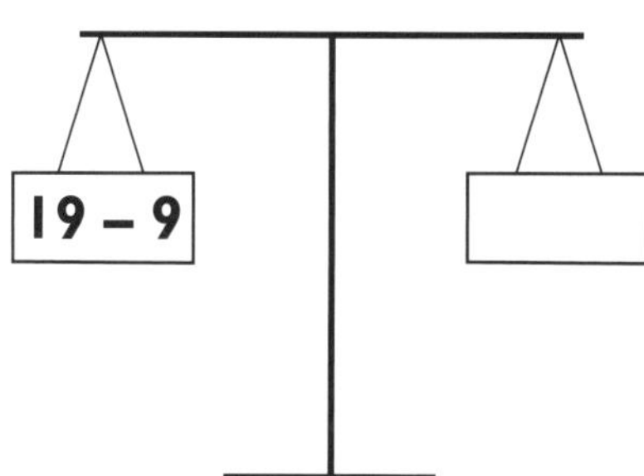

Which other subtraction facts will make the balance level?

Record the number facts suggested by students on the other side of the balance. Emphasise the idea of equality, that is, that 19 – 9 'is the same as' 17 – 7. Repeat the activity for: 18 – 6 and 15 – 5.

Something missing?

Show this number sentence: ______ ______ **+** ______ **=** ______ **9**

Can you work out which digits are missing?

Encourage students to work in pairs to record the possible solutions.

How did you find your solutions? What did you notice? (The tens are the same on both sides. . .) Why do you think this happens? (Because 2 + 7 = 9, 52 + 7 = 59 and 62 + 7 = 69.)

Discuss the students' responses.

Guess and check

Can you guess my number? When I add 5, the answer is 16.

Encourage students to explain how they worked it out. *(16 take away 2 is 14. Take away 2 is 12. It is 11. I imagined a number line and counted back from 16.)*

Invite students to pose their own 'guess my number' problems. Choose student problems to be solved by the rest of the class.

(What is my number? When I take 6 away, the answer is 21.)

MIDDLE – YEARS 3 AND 4

Number and Algebra

NUMBER AND PLACE VALUE

Year 3

Identifying odds and evens

Materials

- a set of number cards
- blank cards
- a black marker

Show these numbers on cards:

12	65	3	82	70	51	93	8	34	100	17	116

Which of these numbers are odd? Which of these numbers are even?

Make two headings (odd and even) and choose students to place the number cards under the correct headings. Looking at how the numbers are grouped, ask the students:

What makes an odd number? What makes a number even? Can all numbers be grouped under these headings?

Call for volunteers to write a number on a card and place it under the odd or even heading.

Making odds and evens

Materials

- a set of number cards from 0 to 9

Show the class a set of number cards from 0 to 9. Ask a student to randomly choose three cards from the deck, for example, **2**, **7**, **4**.

How many different odd numbers can you make with these digits? (27, 47, 247) How do you know the numbers are odd? (The last digit is odd. You cannot divide them into two groups.) How many different even numbers can you make with these digits? How do you know the numbers are even?

Repeat the activity with another three number cards.

Odds and evens pattern

Show a hundreds chart:

1	2	3	4	5	6	7	8	9	10
11	12	13	14	15	16	17	18	19	20
21	22	23	24	25	26	27	28	29	30
31	32	33	34	35	36	37	38	39	40
41	42	43	44	45	46	47	48	49	50
51	52	53	54	55	56	57	58	59	60
61	62	63	64	65	66	67	68	69	70
71	72	73	74	75	76	77	78	79	80
81	82	83	84	85	86	87	88	89	90
91	92	93	94	95	96	97	98	99	100

Starting on 2, skip count by twos to 100. Circle these numbers as the students count.

What pattern can you see? What do we call the numbers circled?

Starting on 1, skip count by twos to 99. Underline these numbers as the students count.

What pattern can you see this time? What do we call the numbers underlined? After 100, what would be the next ten even and odd numbers?

Letterbox numbers

If the letterboxes on one side of the street show the numbers ***298***, ***300***, ***302***, ***304***, ***306*** *and* ***308***, *what letterbox numbers would be on the other side of the street? (297, 299, 301, 303, 305, 307, 309) What number is on your letterbox? What number would be opposite your house?*

Variation

Investigate what happens to letterbox/house numbers in a court or on different types of streets.

Representing numbers (1)

Materials

- a class set of calculators

How can I make the calculator show ***65***, *using the* [+] *key? (60 + 5)*

Discuss different students' responses.

Can we represent other 2-digit numbers like this? How would we represent 89?

Invite individual students to enter their responses on the calculator. Repeat for other 2-digit numbers.

Representing numbers (2)

Materials

- a class set of calculators

Can you make the calculator show ***271*** *using the* [+] *key? (200 + 70 + 1. 200 + 71. 270 + 1)*

Discuss different student responses such as 200 + 71, 270 + 1 . . .

Can we show other 3-digit numbers like this? How would we represent 625?

Invite individual students to enter their responses on the calculator. Repeat for other 3-digit numbers.

Line jumps (1)

Show this number line segment:

0 **200**

Can you get from 0 to 200 in two jumps? What would those jumps look like? Which numbers would you land on? (0, 50, 100; 0, 100, 200; 0, 150, 200, etc.)

Discuss the range of possible answers. Show the jumps suggested by students on the number line and discuss the different solutions. Repeat the activity using three jumps – they do not have to be of equal size.

Line jumps (2)

Show this number line segment:

0 **500**

Can you get from 0 to 500 in two jumps? What would those jumps look like? Which numbers would you land on? (0, 200, 500: 0, 100, 500; 0, 250, 500, etc.)

Discuss the range of possible answers. Show the jumps suggested by students on the number line and discuss the different solutions. Repeat the activity using five jumps – they do not have to be of equal size.

Equal jumps

Show this number line segment:

Can you get from 0 to 1000 in two equal jumps? Which numbers would you land on? (0, 500, 1000 . . .)

Show the jumps suggested by students on the number line and discuss the different solutions. (In this case, there is only one possible answer.)

Can you get from 0 to 1000 in ten equal jumps? What numbers would you land on? (0, 100, 200 . . . 800, 900, 1000)

Show the jumps suggested by students on the number line and discuss the different solutions. Count to 1000, forwards and backwards by hundreds, using the number line segment.

Extension

Repeat the activity by using five jumps and four jumps.

One before and one after

Materials

- a class set of calculators

Enter [1] [0] [0] on the calculator.

What is the number that is one more than 100? (101, 100 + 1) How can we check the answer?

Enter [+] [1] [=] and then clear the calculator and re-enter [1] [0] [0].

What is the number that is one less than 100? (99, 100 – 1) How can we check the answer?

Enter [–] [1] [=]. Repeat the activity for other 3-digit numbers.

Extension

Discuss why 99 is not written as 909.

Ten before and ten after

Materials

- a class set of calculators

Enter [2] [6] [8] on the calculator.

What is the number that is 10 more than 268? (278, 268 + 10) How can we check the answer?

Enter [+] [1] [0] [=] and then clear the calculator and re-enter [2] [6] [8], or enter [–] [1] [0] [=].

What is the number that is 10 less than 268? (258, 268 – 10) How can we check the answer?

Enter [–] [1] [0] [=]. Repeat the activity for other 3-digit numbers. Extend to numbers that are 100 more or less than a particular 3-digit number.

Guess, check and refine

Show this number line segment:

I am thinking of a number between 100 and 200. Can you guess what it is? (168, 121, 199 . . .)

Allow the students to take turns guessing the number. Show each guess on the number line. Respond to students' guesses until they arrive at the selected number.

(My number is less than 168 . . . . My number is more than 168 . . .)

Repeat the activity by choosing a number in a different range, such as 300 to 400, 260 to 360, 225 to 325.

Variation

Students can take turns to choose numbers that their classmates try to guess and show these guesses on the number line.

Extension

Gradually increase the range of numbers, such as 200 to 400, 100 to 500, and 1 to 1000.

Higher or lower?

I am thinking of a 3-digit number between 400 and 500. Can you guess what it is? (416, 445 . . . 481 . . .)

Allow the students to take turns at guessing the number. Respond to students' suggestions with 'higher' or 'lower' until the correct number is reached.

Variation

Repeat this activity often and set a limit of 20 questions if desired. This activity is diagnostic of students' sense of the relative size of numbers and their ability to retain an idea and use reasoning to arrive at a solution. Students who are not confident in participating may need further experience with manipulative materials.

Less than or greater than (2 digits)

Show these numbers: **53, 35**

Which of these two numbers is the greater? (53) How do you know? (It is the one with the highest digit in the tens place.)

Discus this statement: 53 is greater than 35.

Does anyone know a simpler way to write this?

Show the statement using the symbol >, explaining that the 'jaws' of the symbol always eat the larger number.

Which of these two numbers is the lesser?

Discuss this statement: 35 is less than 53. Repeat the activity, inserting the 'less than' symbol, <.

Show these number pairs and discuss the correct symbol for each: **34** ______ **43,**

71 ______ **68, 44** ______ **61.**

Digit combo (1)

Show these numbers: **2, 5, 7**

What is the largest number you can make with these digits? (752) Why did you start with the 7? (Place value, 7 is the largest number in the hundreds column.) How could you arrange these digits to make the smallest number possible? (257) Why is the smallest number not 275? (Place value, 5 tens is less than 7 tens.)

Digit combo (2)

Show these numbers: **0, 9, 3**

What is the largest number you can make by rearranging these digits? (930) Why did you start with the 9? (Place value, 9 hundreds is greater than 3 hundreds.) How will you arrange these digits to make the smallest number possible? (309) Why is the smallest number not 039?

Explain that zero cannot be used as a place holder in the first column to the left, which is the hundreds column here. So, 039 would really be 39 so would not include all three numbers. If necessary for the discussion, note that there are some instances of using zero in the first column to the left, when writing the date, a telephone number or for a credit card or PIN etc.

Break it down (1)

Show this number: **200**

How many hundreds in 200? How many tens in 200? How many ones in 200? (2, 20, 200)

Students will often respond incorrectly by giving the place value of each digit in the number. So, it may be necessary to structure the different representations of this number with manipulative materials so that students develop more complete mental pictures and understandings.

Can you picture 200 ones? Can you picture 20 tens and 9 ones? Can you picture 2 hundreds? What do they look like?

Discuss the students' mental imagery of these numbers.

Break it down (2)

Show this number: **379**

How many hundreds in 379? How many tens in 379? How many ones in 379? (3, 37, 379) Can you picture 379 ones? Can you picture 37 tens and 9 ones? Can you picture 3 hundreds, 7 tens and 9 ones? What do they look like?

Discuss the students' mental imagery of these numbers.

Extension

Repeat the activity for 405 and 420.

Expanding numbers (1)

Show this problem: **25 = ______ tens + ______ ones**

What is the place value of the 2 in 25? (2 tens) What is the place value of the 5 in 25? (5 ones)

Can you picture the size of the 2 tens? Can you picture the size of the 5 ones? What do they look like?

Discuss the students' mental imagery of these numbers. It might be useful to show a concrete representation of the number using craft sticks, ten-frames, Unifix or Base 10 materials so that students develop flexible understandings. Encourage students to form mental images of numbers by directed language (e.g. visualise, imagine, picture) at all times.

Variation

Repeat the activity for other 2-digit numbers. Represent the numbers using manipulative materials.

Expanding numbers (2)

Show this problem: **48 = ______ tens + ______ ones**

How many tens are there in 48? (4) If there are 4 tens in 48, how many ones are there? (8) Can you picture 48 as ones? How many ones are there in 48? (48)

Variation

Repeat the activity for other 2-digit numbers.

Expanding numbers (3)

Show this problem: **263 = ______ hundreds + ______ tens + ______ ones**

What is the place value of the 2? (2 hundreds) What is the place value of the 6? (6 tens) What is the place value of the 3? (3 ones) Can you picture the size of the 2 hundreds? Can you picture the size of the 6 tens? Can you picture the size of the 3 ones? What do they look like?

Discuss the students' mental imagery of these numbers.

Variation

Repeat the activity for other 3-digit numbers. Represent the numbers using manipulative materials.

Year 4

Identifying odds and evens

Show these numbers: **358, 2096, 501, 892, 4316, 1999, 3465, 78, 5621, 8314, $2.65, 54.82**

In what ways could we group these numbers? (odd and even, numbers with or without a decimal, numbers under or over 100 or 1000, numbers or money)

Make up lists and ask for volunteers to write each number under its correct heading. Ask for students to make a few more suggestions for each heading.

What would happen to the odd and even numbers if we were to add 1 to each number?

Discuss how an odd number would become even and an even number would become odd.

Odds and evens

Materials

- a set of number cards from 0 to 9

Show these number cards:

6 **5** **3** **8**

How many different odd numbers can you make with these digits? (12 numbers: 8653, 8635, 8563, 8365, 6853, 6835, 6385, 6583, 5863, 5683, 3685, 3865) How many different even numbers can you make with these digits? (12 numbers: 8536, 8356, 6538, 6358, 5836, 5638, 5368, 5386, 3568, 3586, 3658, 3856)

Broaden the discussion to include other number questions.

How many numbers can you make that are less than 5000? (6 numbers: 3568, 3586, 3685, 3658, 3856, 3865) How many numbers can you make that are greater than 4000 but less than 5000? (None) How many numbers can you make that are divisible by 5? (6 numbers: 8635, 8365, 6835, 6385, 3685, 3865)

Ask the students to devise their own questions based on the 4-digit numbers above, or other 4-digit numbers.

Add and subtract odds and evens

Show these number sentences: **12 + 6, 17 + 5, 8 + 4, 8 + 7, 11 + 5, 3 + 9**

What happens when you add an even number with an even number? (Even + even = even) What happens when you add two odd numbers together? (Odd + odd = even) What happens when you add an odd number and an even number? (Even + odd = odd)

Have students list examples in pairs or small groups.

Do the same rules apply with subtraction? (Yes, even – even = even, odd – odd = even, even – odd = odd)

Allow students time to discuss and provide examples for subtraction.

Multiplying odds and evens

Materials

- a multiplication chart

Show a multiplication chart:

1	2	3	4	5	6	7	8	9	10	11	12
2											
3											
4											
5											
6											
7											
8											
9											
10											
11											
12											

What multiples of numbers count by even numbers only? (twos, fours, sixes, eights, tens, twelves)

Are there any multiples of numbers that count by odd numbers only? (No) What multiples of numbers count by odd and even numbers? (Ones, threes, fives, sevens, nines, elevens)

Discuss the students' suggestions.

What happens when you multiply an even number with an even number? (The answer is always even.) What happens when you multiply an odd number with an odd numnumber with an odd number? (The answer is always even.)

Discuss and make a list of examples for each kind of multiplication.

Three-digit puzzle

Show this number puzzle:

Can you find the missing numbers? How did you work out your answer? (There's a pattern.

You count along by ones and down by tens.)

Repeat the activity for different shaped puzzles and other 3-digit numbers.

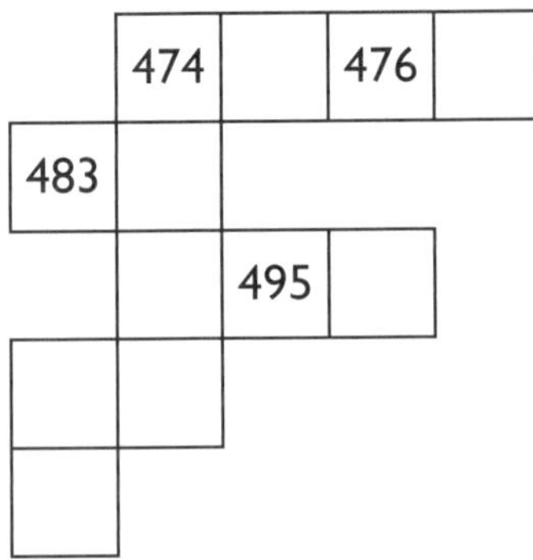

Four-digit puzzle

Show this number puzzle:

Can you find the missing numbers? How did you work out your answer? (There's a pattern. The number underneath is 10 more. When you count along, it goes by ones.)

Repeat the activity for different shaped pieces and other 4-digit numbers.

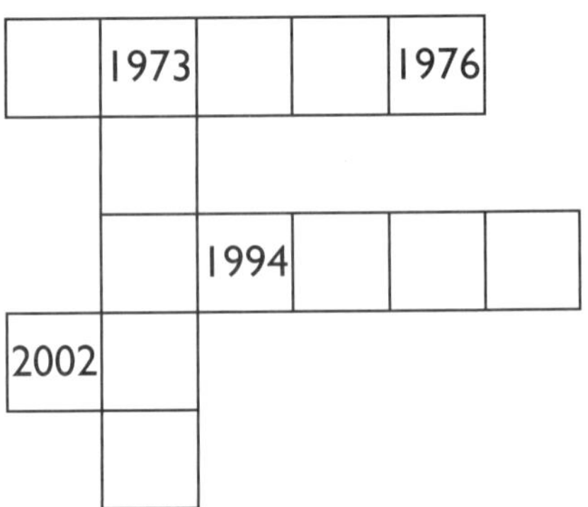

Number line jumps

Show an open number line segment from 1000 to 2000, with intervals of 100:

1100 1200 1300 1400 1500 1600 1700 1800 1900

1000 **2000**

How many ways are there to get from 1000 to 2000 using three jumps? What would those jumps look like?

Accept a wide range of answers. Show the jumps suggested by students on the number line. Repeat the activity using four jumps and other intervals, for example, 3000 to 4000.

Equal jumps

Show an open number line segment from 1000 to 2000, with intervals of 100:

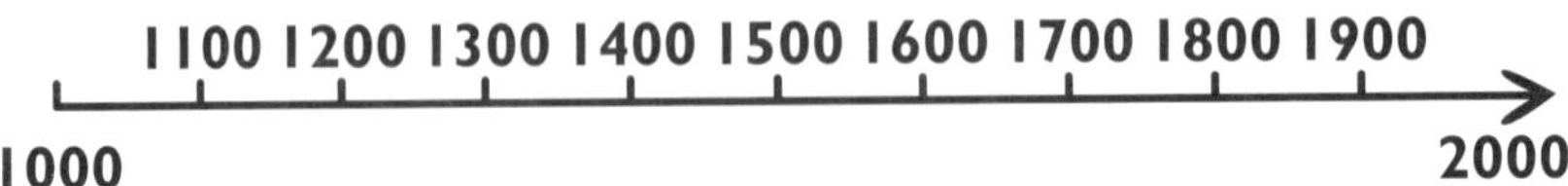

Can you get from 1000 to 2000 in two equal jumps? What would those jumps look like? (land on 1500)

Show the jumps suggested by the students on the number line.

Can you get from 1000 to 2000 in five equal jumps? (Land on 1200, 1400, 1600, 1800) Can you get from 1000 to 2000 in four equal jumps? (Land on 1250, 1500, 1750) What would those jumps look like?

Show the jumps suggested by the students on the number line. As a class, practise counting along the number line forwards and backwards by 500s, 200s and 250s.

Hundreds jumps

Show an open number line segment from 1000 to 2000, with intervals of 100:

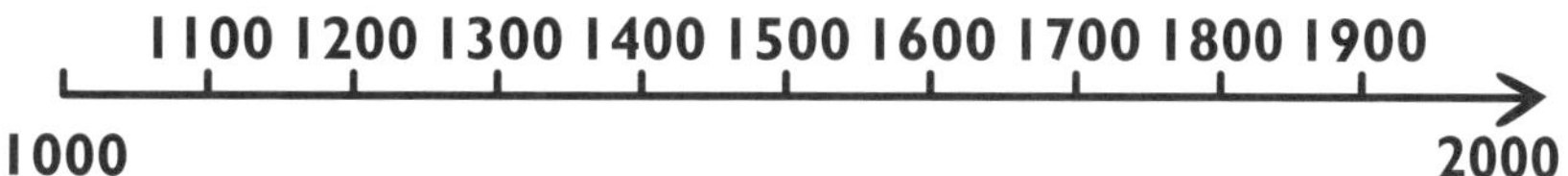

Invite the class to count forwards and backwards by 100 from 1000 to 2000.

If I started at 1000, how many jumps of 100 would it take to get to 1200? (Two jumps) If I was at 1200 and made three more jumps of 100, where would I be? (1500) How many jumps of 100 would get me from 1500 to 2000? (Five jumps)

Show the jumps suggested by the students on the number line.

Decade jumps

Show an open number line segment from 1000 to 1500, with intervals of 10. One section is shown here:

Can you count from 1000 to 1200 by tens? Can you count from 1500 back to 1300 by tens? Can you count from 1225 to 1425 by tens? Can you count from 1495 back to 1295 by tens?

Repeat the activity. Ask students to choose a starting number and a suitable number to count by. Include numbers on and off the decade.

Guess, check and refine

Show an open number line segment from 3000 to 4000, with intervals of 100:

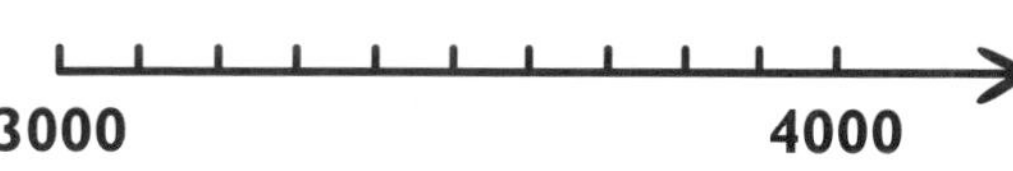

Choose a number in this range without revealing it to the students.

I am thinking of a number between 3400 and 3700. Can you guess my number?

Allow the students in turn to guess the number.

(Is it between 3010 and 3020? Is the number larger than 3450?) No, the number is not larger than 3450.

Mark each suggested number on the number line segment until the selected number is reached.

Variation

Play the game within different ranges. Allow students to write their number for increased motivation.

Imagining numbers

Materials

- Base 10 materials

Show this number description:

5 thousands 4 hundreds 2 tens and 6 ones

Can you imagine what this number looks like using Base 10 materials?

Ask students to describe their representation of the number. Show the Base 10 materials. Show the numerals for this number: 5426

Repeat the activity for these number descriptions:

4 thousands 2 hundreds 6 tens and 1 ones

3 thousands 9 hundreds 8 tens and 4 ones

6 thousands 1 hundreds 0 tens and 6 ones

Three-digit breakdown

Materials

- Base 10 materials

Show this number: **654**

How many hundreds in 654? (6 hundreds) How many tens in 654? (65 tens) How many ones in 654? (654 ones)

Note that some students may say there are 5 tens in 654. Explain that 5 tens is the *value* of the tens place but not the *total number* of tens.

Can you picture 654 ones? Can you picture 65 tens and 4 ones? Can you picture 6 hundreds, 5 tens and 4 ones?

Discuss the students' mental imagery of these numbers. Use the Base 10 materials to demonstrate the different representations of 654.

Four-digit breakdown

Materials

- Base 10 materials

Show this number: **2379**

How many thousands in 2379? (2 thousands) How many hundreds in 2379? (23 hundreds) How many tens in 2379? (237 tens) How many ones in 2379? (2379 ones) Can you picture 2379 ones? Can you picture 23 hundreds and 79 ones? Can you picture 237 tens and 9 ones?

Discuss the students' mental imagery of these numbers. Use the Base 10 materials to demonstrate the different representations of 2379.

Numeral shuffle

Materials

- a set of number cards from 0 to 9

Show the class the set of number cards. Select four students to choose one number card each.

How could you arrange the numeral cards to make the largest number possible? (Thousands are bigger, so you put the largest number in that place.) Why did you arrange the cards that way? What would be the smallest number possible? Why did you place the smallest number in the thousands place? Can zero be used to begin your number? (No, zero holds the place in the middle or at the end of the number only.)

Repeat the activity with another group of students.

Second highest

Materials

- a set of number cards from 0 to 9

Show these number cards:

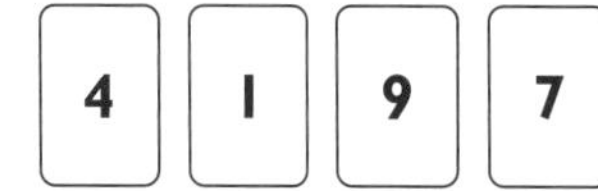

What is the highest number that you can make with these cards? (9741) What would be the second-highest number you could make? (If you change the ones and tens you get 9714. That is closest to 9741.) Why is the second-largest number not 7941? (7000 is much smaller than 9000.)

Variation

Repeat this activity with four different cards to identify lowest and second-lowest 4-digit numbers.

Digit show

Materials

- two sets of number cards from 0 to 9
- two place value charts

Thousands	Hundreds	Tens	Ones

Divide the class into two teams. Explain to the students that you are playing a game where the aim is to make the largest 4-digit number possible. Choose one student from each team in turn, to select a number card. Students then place the card in the column of their choice on their team's place value chart. After each student in the team has placed their card, the team with the highest number is the winner.

Variation

Repeat the game with the aim of making the smallest 4-digit number possible.

Order for five

Show these numbers on separate cards:

Choose five students to choose one card each.

792	762	1040	1004	961

Can you order these numbers from lowest to highest? (762, 792, 961, 1004, 1040) What is the second highest number in the sequence? How do you know? (1040 is higher than 1004 because the 4 is in the tens place, 40 is larger than 4.) Which number is in the middle? (It is 961. There are two numbers below it and two numbers above it.)

Tens order

Show these numbers on separate cards:

Choose five students to choose one card each.

1367	1376	1370	1307	1360

Can you order these numbers from smallest to largest? (1307, 1360, 1367, 1370, 1376) How do you know that 1376 is the largest number? (1376 is the largest as it has 7 tens in the tens place.)

Discuss the relative values of each number in terms of its place value. With these numbers, the relative sizes of the tens determine their order.

Higher or lower?

Materials

- a collection of 4-digit numbers written on cardboard strips

Ask a student to choose one of the strips and to keep the number secret. The rest of the students take turns to guess the number by suggesting possible numbers. The student with the strip can only respond 'higher' or 'lower' to the suggested numbers until the target is reached.

Variation

Record the students' responses progressively on a number line as the game is played.

Less than or greater than?

Show these pairs of numbers: **2345** and **2354**, **3100** and **3005**, **4102** and **4099**, **8890** and **8989**

Which of these two numbers is the greater? How do you know? (2354 is the greater as it has 5 tens, 2345 has 4 tens.) Which is the lesser? (3100 is greater as it has 31 hundreds, 3005 has 30 hundreds.)

Discuss the relative sizes of the numbers in terms of their place value. Revise the use of the symbols for 'greater than' > and 'less than' <, explaining that the 'jaws' or 'mouth' of the symbol always eats the larger number. Ask the students to insert the correct symbols in between the pairs of numbers.

Spot the largest

Show these series of numbers:

Which is the largest number in each group? (6542, 6000, 1111, 9690) How do you know? (You need to check the thousands and hundreds place first.)

Discuss the relative sizes of the numbers in terms of their place value. Identify the largest number in each series.

3452	**4452**	**6542**	**5542**
5886	**886**	**6000**	**5990**
1110	**1011**	**1101**	**1111**
9609	**9069**	**9690**	**9669**

Spot the smallest

Show these series of numbers:

Which is the smallest number in the third group of numbers? (It is either 8445 or 8455.) How do you know? How do you know? (45 is smaller than 55 so it is 8445.)

Discuss the relative sizes of the numbers in terms of their place value. Identify the smallest number in each series.

3890	**3980**	**3908**	**3981**
9099	**9019**	**9109**	**9009**
8544	**8445**	**8545**	**8455**
2012	**2220**	**2221**	**2022**

Numerals to words and words to numerals

Show these numbers in words:

Four thousand, nine hundred and fifty-seven

Six thousand, three hundred and forty-four

Two thousand, eight hundred

Six thousand and fifty-one

Can you write the numerals for these numbers? (4957, 6344, 2800, 6051)

List the students' responses next to each number. Explain the convention in naming numbers containing zeros: the place value of the column containing zero is not read, for example, 4009 is read as 'four thousand and nine'—not as 'four thousand zero hundreds, zero tens and nine'.

Show these numbers: **4898, 6392, 8620, 1896, 4070**

Can you write these numerals in words?

List the students' responses next to each number.

Addition and subtraction

Year 3

Dice combinations

Materials

- 3 dice

If I rolled three dice and the total was 10, which numbers were showing on the faces of the dice? (6 + 2 + 2 = 10, 5 + 2 + 3 = 10, 4 + 3 + 3 = 10)

Record the combinations of numbers suggested by students to make 10. List all possible combinations and discuss.

Addition facts practice

Show a small hexagon with a 2-digit number inside it:

Can you tell me two numbers that add up to this number? (32 + 32, 60 + 4)
Can you suggest other combinations of two numbers that add up to 64?

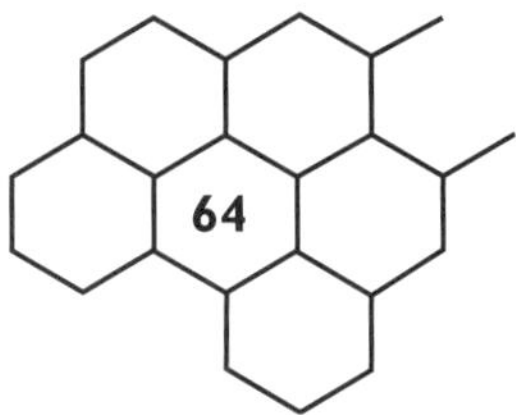

While students are mentally calculating their combinations, show a tessellated pattern of hexagons around each side of the original hexagon. List the students' number facts in the hexagons around the original hexagon.

Addition fact combinations

Show six different 1-digit numbers: **4**, **7**, **5**, **9**, **2**, **8**

Choose any two numbers. Can you work out their total? (4 + 7 = 11, 9 + 8 = 17, 2 + 5 = 7, 7 + 5 =12)
Can you work out the totals for the other combinations of numbers?

Ask students to work in pairs and record all possible totals. Check the answers and discuss. Observe students while they are working. Pay special attention to the different strategies employed by students: those who have memorised number facts, those who use basic strategies such as counting by ones, and so on.

Variation

Repeat the activity, choosing three numbers at a time.

Mixed fact combinations

Show six different 1- and 2-digit numbers: **15**, **6**, **18**, **8**, **2**, **11**

Ask students to work out as many addition and subtraction combinations as they can, using any two of the numbers.

(18 – 6 is 12, 15 + 6 is 21)

Ask students to work in pairs and record all possible totals. Check all possible answers and discuss.

Variation

Repeat the activity using two operations in the same number sentence, for example, **18 – 6 + 8 = 20.**

Number line patterns

Show this number: **40**

Can you tell me two numbers that add up to this number? (10 + 30 = 40, 20 + 20 = 40) Can you list other combinations of two numbers?

While students are working, show a number line from 0 to 40, marking the numbers in between. Show the students' number sentences, using steps on the number line to indicate pairs of numbers that total 40.

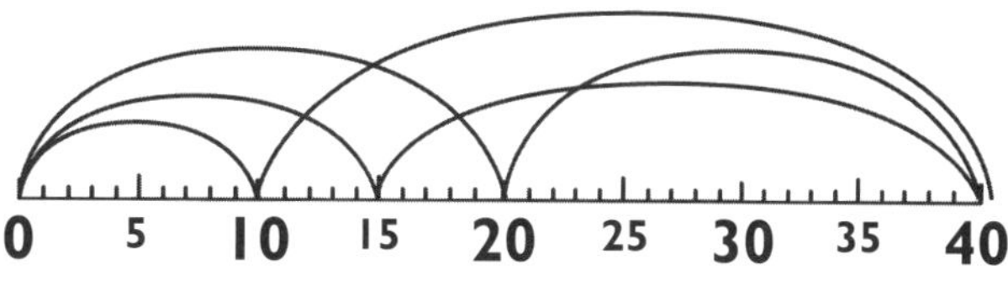

List all the number facts in a table and discuss any patterns formed.

Subtraction facts practice

Show a small hexagon with the number 5 inside it:

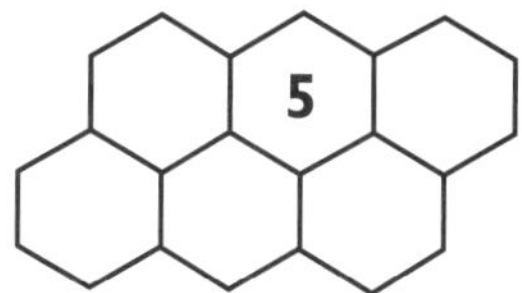

This is the answer to a number sentence using subtraction. Can you suggest number sentences that could have an answer of 5? (14 – 9 = 5, 10 – 5 = 5, 23 – 18 = 5)

While students are mentally calculating their number sentences, show a tessellated pattern of hexagons around each side of the original hexagon. List the students' number facts in the hexagons around the original hexagon. Choose another single-digit number and repeat the activity.

Addition patterns

Materials

- a hundreds chart

Show the number 17 on the hundreds chart.

Can you tell me how to add 10 to 17 using the chart? (To add 10, you go down one square.)

Discuss the pattern with the students.

What if I add 12 to 27? What is the pattern there? (It is down 1 square and across 2 squares.)

Discuss the students' responses.

How could you add 42 and 23 on the chart?

1	2	3	4	5	6	7	8	9	10
11	12	13	14	15	16	17	18	19	20
21	22	23	24	25	26	27	28	29	30
31	32	33	34	35	36	37	38	39	40
41	42	43	44	45	46	47	48	49	50
51	52	53	54	55	56	57	58	59	60
61	62	63	64	65	66	67	68	69	70
71	72	73	74	75	76	77	78	79	80
81	82	83	84	85	86	87	88	89	90
91	92	93	94	95	96	97	98	99	100

Discuss the addition patterns that the students suggest.

Subtraction patterns

Materials

- a hundreds chart

Show the number 95 on the hundreds chart.

Can you tell me how to subtract 10 from 95 using the chart? (To subtract 10, you go up one square.)

Discuss the pattern with the students.

What if I subtract 23 from 85? What is the pattern there? (It is up 2 squares and 3 squares left.)

Discuss the students' responses. Ask the students to suggest another 2-digit number that could be subtracted from 62. Repeat the activity, discussing the subtraction patterns that the students suggest.

Two-digit near doubles

Show this number sentence: **70 + 72 =** ______

Can you tell me how you worked it out? (70 + 70 + 2 = 142)

Discuss the different strategies used by students. Focus on the near doubles strategy (70 + 70 + 2 = 142).

Repeat the activity for these number sentences: **40 + 41 =** ______, **50 + 52 =** ______, **63 + 60 =** ______, **73 + 70 =** ______.

Note that focusing on a particular strategy in discussions with students is not intended to prescribe a certain approach. The intention is to develop number sense by modelling a wide range of mental computation strategies from which they can choose.

Three-digit doubles

Show this number sentence: **400 + 400 =** ______

Can you tell me how you worked it out? (4 + 4 = 8 × 100 = 800, 800)

Discuss the different strategies used by students.

Did anyone use a pattern to help work it out?

Show the pattern: 4 + 4 = 8, 40 + 40 = 80, 400 + 400 = 800, 4000 + 4000 = 8000.

Variation

Ask the students to suggest and solve other 3-digit doubles.

Three-digit near doubles

Show this number sentence: **120 + 119 =** ______

Can you tell me how you worked it out? (239, 240 – 1, 120 + 120 – 1)

Discuss the different strategies used by students. Focus on the near doubles strategy (120 + 120 – 1 = 239).

Repeat the activity for these number sentences: **150 + 149**, **240 + 239**, **320 + 319**.

Before calling for answers that involve mental computation, always allow students enough time for mathematical thinking. It is useful to list all responses without comment at first, including those which may be incorrect. Each strategy can then be reviewed.

Hidden numbers

Materials

- ten-strips

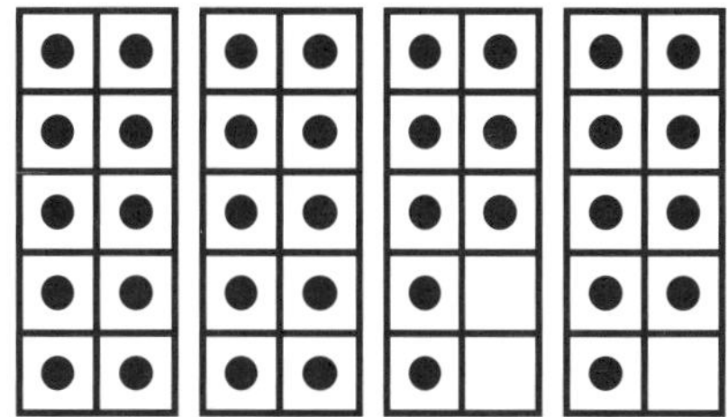

Show 2 tens, 8 ones and 9 ones, then and cover the 9 ones.

What number is shown here? (20 + 8, 28) If I had 37 altogether, how many ones are hidden? (I can count backwards: 37, 36 . . ., I can count on 28, 29 . . ., I can take away 10 and add 1 . . ., that is 9.)

Discuss the students' strategies.

Subtraction combinations

Show one circle containing the numbers 12, 14 and 19 and one circle containing the numbers 6, 7 and 9. Ask students to choose any number from the second circle and subtract it from a number in the first circle.

Can you work out the answers for all the other subtractions?

Have students work in pairs and record all possible answers. Check their answers and discuss their strategies.

Changing order

Show these number sentences:

15 + 7 + 5 = ______ **21 + 6 + 9 =** ______ **60 + 8 + 40 =** ______

Can you see a quick way to add these numbers? (Do the 15 + 5 first. That is 20. Add 7 which makes 27, 60 + 40 is 100, 8 more is 108.)

Explain to students that the numbers can be added in any order. Have the students suggest some similar number sentences of their own.

Jump or split?

Show this number sentence: **35 + 24 =** ______

How would you solve this problem?

Discuss the strategies that students provide, such as the jump strategy (35 + 20 = 55, 55 + 4 = 59) or the split strategy (30 + 20 = 50, 50 + 5 + 4 = 59). Repeat the process for another number sentence: **42 + 26 =** ______.

Once again, it is not the intention to prescribe certain mental strategies. Number sense is developed by making students aware of a wide range of strategies that they can use during mental computation.

Larger jumps and splits

Show this number sentence: **43 + 65 =** ______

How would you solve this problem?

Discuss the strategies that students provide, such as the jump strategy (43 + 60 = 103, 103 + 5 = 108) or the split strategy (40 + 60 = 100, 100 + 3 + 5 = 108). Repeat the process for another number sentence: **51 + 74 =** ______

Ask students to use a different solution strategy for this problem.

Compensation strategy

Show this number sentence: **63 + 29 =** ______

How would you solve this problem?

Discuss the range of strategies that students provide. Introduce the compensation strategy if the students have not already suggested it (63 + 29; 63 + 30 = 93 – 1 = 92).

Repeat the process for another number sentence: **54 + 39 =** ______

Explain that the compensation strategy is useful when one of the numbers is close to a multiple of 10, for example, 63 + 29 = 63 + 30 – 1 = 92. Encourage students to always scan the numbers in any number sentence and look for their characteristics: close to multiple of ten, doubles, or near doubles, and so on.

Make and record

Materials

- Base 10 materials

There were 254 students in a school but 85 left to attend a swimming carnival. How many students stayed behind at school?

Can you work out this problem using the Base 10 materials?

Encourage students to develop and discuss a variety of solutions using the Base 10 materials. Note that the standard algorithmic method involving decomposition (trading) does not necessarily have to be used at this stage.

Show and record

There were 552 students in a school. On sports day, 326 students stayed behind at school and the rest travelled to a sports ground by bus. How many students went to the sports ground? Can you work out this problem using a number line?

Discuss the different counting strategies used by the students.

Three for 350

Materials

- Unifix or Multilink blocks
- an abacus
- Base 10 materials

Show this number sentence: ______ **+** ______ **+** ______ **= 350**

Can you tell me three numbers that add up to this number? (200 + 100 + 50 = 350, 150 + 150 + 50 = 350)

Ask pairs of students to choose one number sentence, then represent and explain the solution with manipulatives. Show and discuss the students' representations with the whole class. Students could also record their workings using diagrams and symbols. Note that this will help them in moving towards the final algorithm.

Three for 500

Materials

- paper
- coloured pencils

Show this number sentence: ______ **+** ______ **+** ______ **= 500**

Can you tell me three numbers that add up to this number?

Ask each student to choose a number sentence, then represent and explain their solution on paper using diagrams and symbols. Choose several students to share their workings with the class.

Mixed fact strategies

Show these number sentences: **18 – 9 =** ______, **14 – 6 =** ______, **16 + 9 =** ______, **17 + 5 =** ______

Can you tell me how to work out these problems? (14 – 4 = 10, take away 2 more, you get 8, 9 + 9 = 18 so 19 – 9 = 9, 17 + 3 = 20, add 2 more = 22)

Discuss the various strategies used by students to solve the problems.

Can you make up some more problems using these strategies?

Discuss the students' suggestions and ask them to justify their solutions. Note that asking students to justify their approach during reporting and sharing sessions is crucial for the development of individual mathematical thinking.

Near decade strategies

Show these number sentences: **18 + 5 =** ______, **13 + 14 =** ______, **32 – 4 =** ______, **24 – 12 =** ______

Can you tell me how to work out these problems? (32 – 2 is 30, take away 2 more and it is 28. 18 + 2 is 20, add 3 more and it is 23.)

Discuss the various strategies used by students to solve the problems.

Can you make up some more problems using these strategies?

Discuss the students' suggestions and ask them to justify their solutions.

Inverse operations

Show these number sentences: **16 – 8 = 8, 24 – 9 = 15, 32 – 12 = 20**

How can you check these subtraction problems using addition? (8 + 8 = 16, 15 + 9 = 24, 20 + 12 = 32)

Show these number sentences: **12 + 9 = 21, 18 + 7 = 25, 29 + 8 = 37**

How can you check these addition problems using subtraction? (21 – 9 = 12, 21 – 12 = 9, 25 – 7 = 18, 25 – 18 = 7, 37 – 8 = 29, 37 – 29 = 8)

Remind students that addition and subtraction are inverse operations, which means subtraction problems can be checked by doing addition and addition can be checked by doing subtraction. Ask students to demonstrate more number problems that can be solved by using inverse operations.

Backtracking

Show this number: **24**

How many addition number sentences can you make that equal 24? How can we check they are correct using subtraction? (16 + 8 = 24; therefore, 24 – 8 = 16, 15 + 9 = 24 and 24 – 9 = 15)

Discuss how this method of using inverse operations is also called backtracking. Ask students to suggest more numbers and number sentences that can demonstrate backtracking.

Year 4

Number strip addition

Show this number strip:

0	1	2	3	4	5	6	7	8	9	10

Can you tell me three numbers from the strip that add up to 10? (6 and 1 and 3 equals 10. 2 and 3 and 5 is 10) Can you find all the groups of three numbers that add up to 10? Is there a pattern?

Discuss all the answers suggested by the students.

0 + 1 + 9 = 10	**1 + 2 + 7 = 10**	**2 + 3 + 5 = 10**
0 + 2 + 8 =10	**1 + 3 + 6 =10**	**2 + 4 + 4 = 10**
0 + 3 + 7 = 10	**1 + 4 + 5 =10**	**3 + 5 + 2 = 10**

Does the order in which the numbers are added make any difference to the total?

Two addends to 50

How many pairs of numbers can you find that add up to 50? (25 and 25 equals 50. 10 and 40 add up to 50.)

Can you find a pattern?

1 + 9 **2 + 48** **3 + 47** **4 + 46...**

40 + 10 **41 + 9** **42 + 8** **43 + 7...**

23 + 27 **24 + 26** **25 + 25** **26 + 24** **27 + 23...**

Discuss the students' responses. Before calling for answers that involve mental computation, always allow students enough time for mathematical thinking. It is often useful to list all responses without comment at first, including those that may be incorrect. Each strategy can then be reviewed.

Three addends to 50

How many groups of three numbers can you find that add to 50? (10 and 10 and 30 add to 50; 20 and 20 and 10 equals 50)

Can you find a pattern?

40 + 9 + 1 **39 + 10 + 1** **38 + 11 + 1** **37 + 12 + 1...**

20 + 20 + 10 **20 + 19 + 11** **20 + 18 + 12** **20 + 17 + 13...**

Discuss the students' responses.

Fill the blanks

Show this place value grid:

Th	H	T	O	Total
2000	300	50	6	
	500	70	4	4574
3000	400		5	3495
1000				1986

Can you fill in the blanks across each row? Will you add or subtract? (1986 subtract 1000 is 986. 9 hundreds, 8 tens and 6 units.)

Discuss the strategies suggested by the students.

Two-digit no regrouping

Show this number sentence: **57 – 35 =** ______

What is the answer to this number sentence? Can you tell me how you worked it out? (You can add 5 to both. That is 62 take away 40 leaves 22. You subtract the tens and then the ones. That is 20 and 2 equals 22.)

Discuss and review the different strategies used by students. Encourage as wide a range of strategies as possible.

Two-digit regrouping

Show this number sentence: **47 + 36 =** ______

What is the answer to this number sentence? Can you tell me how you worked it out? (You add the tens and then the ones. That is 7 tens and 13 ones. That equals 83. Add 3 onto 47 and take 3 off 36, that is 50 add 33, which equals 83. You can add 3 to 47 to make it 50, 50 and 36 is 86, take away 3 gives 83.)

Discuss the different strategies used by students. Encourage as wide a range of strategies as possible.

On the level addition

Show this number sentence: **49 + 51 =** ______

What is the answer to this number sentence? Can you tell me how you worked it out?

Discuss the different strategies used by students. Focus on the levelling strategy: *(You add the tens and then the ones. Add 1 onto 49 and take 1 off 51, that is 50 plus 50, which equals 100. You can add 1 to 49 to make it 50. 50 and 51 is 101, take away 1 gives 83.)*

Repeat the activity for these number sentences:

38 + 42 = ______ **48 + 58 =** ______ **28 + 35 =** ______

Note that in focusing on a particular strategy, it is not the intention to prescribe a fixed approach. The intention is to develop number sense by modelling a range of mental strategies that students can use.

On the level subtraction

Show this number sentence: **62 – 38 =** ______

What is the answer to this number sentence? Can you tell me how you worked it out?

Discuss the different strategies used by students. Focus on the levelling strategy: 62 – 38 = 62 + 2 – 38 + 2 = 64 – 40 = 24

Repeat the activity for these number sentences:

52 – 24 = ______ **44 – 27 =** ______ **71 – 26 =** ______

Two-digit compatibles

Show these number sentences: **65 + 28 =** ______, **70 – 39 +** ______

Discuss the students' responses. Focus on the strategy of using a compatible number to obtain the solution.

(I will add 2 to 28 to make it 30. 65 plus 30 equals 95. Take away 2 is 93. 70 take away 40 is 30. Add 1 back on equals 31.)

Repeat the activity for: 55 + 19 = ______, 93 – 29 = ______

Bridging 100

Show these number sentences: **80 + 32 =** ______, **106 – 15 =** ______

What are the answers to these number sentences? Can you tell me how you worked it out? (80 and 20 makes 100. Add 12 more equals 112. If you take 6 off 106 you get 100. Take off 9 more is 91.)

Repeat the activity for: 75 + 27 = ______, 124 – 28 = ______.

Addition fact combinations

Show these numbers: **35**, **13**, **24**, **42**

If you choose any two of these numbers, can you calculate their total? (35 + 13. That is 35 + 10 + 3.) Can you work out the totals for all the other combinations?

Ask students to work in pairs and record all possible totals. Check all possible answers and discuss.

Variation

Repeat the activity with numbers that involve regrouping, for example, 37, 16, 57 and 28.

Extension

Extend the list to five or six numbers or choose three numbers at a time to add.

Mixed fact combinations

Show these numbers: **36**, **45**, **96**, **41**

If you choose any two of these numbers to add or subtract, can you calculate the total? (96 – 45. That is 96 – 40 – 5)

Ask students to work in pairs to find as many addition and subtraction combinations as possible, using any two of the numbers. Check all possible answers and discuss.

Variation

Repeat the activity using two operations in the same number sentence, for example, **45 – 41 + 96 = 100**.

Subtraction hexagons

Show a small hexagon with a 2-digit number inside it:

This is the answer to a subtraction number sentence. Can you tell me what it is?

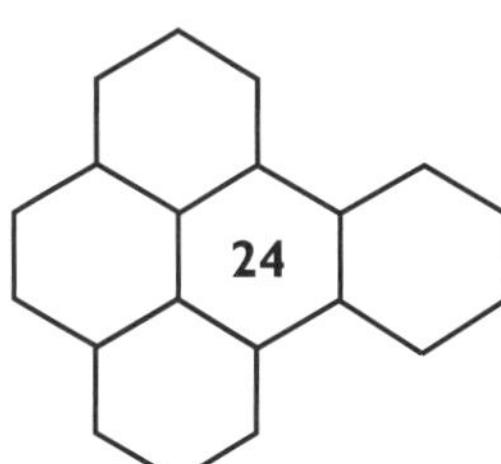

While students are thinking, show a tessellated pattern of hexagons around each side of the original hexagon. List the students' number sentences in the hexagons around the original hexagon.

Variation

Repeat the activity with other 2- or 3-digit numbers.

Three-digit near doubles

Show this number sentence: **249 + 249 =** ______

What is the answer to this? Can you tell me how you worked it out?

Discuss the different strategies used by students. Focus on the strategy of 250 + 250 – 2.

Repeat the activity for: **299 + 299 =** ______, **151 + 151 =** ______.

Number line patterns

Show this number: **470**

Can you tell me two numbers that add up to this number? Can you suggest other combinations?

Show a number line from 0 to 500, with intervals of 10. This is one section:

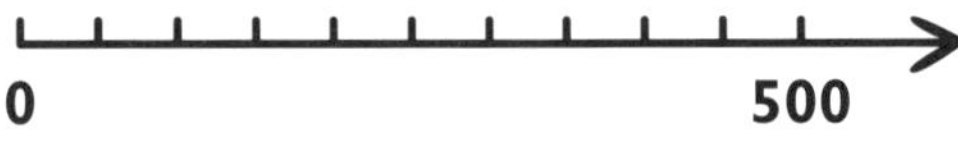

Show loops on the number line to illustrate students' answers, for example, 230 + 240 = 470.

Subtraction combinations

Show one circle containing the numbers 52, 76 and 141 and show one circle containing the numbers 42, 70 and 25. Ask students to choose any number from the second circle and subtract it from the number in the first circle. Discuss the students' responses.

Can you work out the totals for all the other combinations?

Ask students to work in pairs and record all possible answers. Discuss their strategies.

Multiplication and division

Year 3

Symbol meanings

Show the multiplication number sentence: **4 × 3**

What does this mean? (That is 4 times 3. That is 4 groups of 3. That is 4 multiplied by 3.) Can you imagine what 4 groups of 3 looks like?

Encourage students to close their eyes and visualise the groups. Call for volunteers to show arrays. Repeat the activity for: 5 × 4.

Making rectangles

Materials

- counters

Show 24 counters arranged in a rectangle:

How many counters do you see altogether? (There are 24. It is 8 threes.)

How many different rectangles can we make with 24 counters. How many counters long and wide is each rectangle? (24 in 4 rows of 6. And 24 divided into 2 rows, that is 12 in each row.)

Make or show the different rectangles that students suggest. Discuss the different multiplication and division facts that they represent.

Variation

Repeat the activity for numbers such as 36, 45 and 54.

Matter of facts

- counters

Arrange 10 counters in a 5 × 2 array.

How many counters do you see altogether? Can you tell me some grouping and sharing facts for the number 10? (5 twos are 10, 2 fives are 10, 10 divided by 2 is 5, 10 shared between 5 is 2 each.)

List the multiplication and division facts for 10. Repeat the activity with another array, listing multiplication and division facts for the total number of counters.

Rectangle game

Materials

- counters
- Base 10 shorts
- dot papers

This game is an early introduction to prime numbers. Composite numbers can be arranged in rectangular or square arrays with different factors. Prime numbers cannot. Explain the rules of the rectangle game as follows:

Rules

1. Play with a partner.
2. Suggest a number between 1 and 40.
3. If your partner can make a rectangle for that number, they score a point. If they cannot make a rectangle for that number, you get the point.
4. The person with the highest score after five turns each wins.

Variation

Play the game with numbers from 40 to 80 or 50 to 100.

Calculator skip counts

Materials

- a class set of calculators
- scrap paper
- pencils

Do you know how to make the calculator skip count by threes?

Enter [3] [+] [+] and [=] on the calculator. Continue to press [=] so that the multiples of 3 appear. Ask the students to count along as they change the numbers on their display.

(One group of 3 is 3, two groups of 3 are 6, 3 groups of 3 are 9 . . .)

Ask the students to write down the first ten multiples of 3.

Variation

Repeat the activity for multiples of 4 and 6.

Commutative strategy

Show this number sentence: **9 × 3 = 27**

Can you use this number sentence to work out 3 × 9? (Yes, the answer is the same both ways.)

How can you prove your answer?

Students can use counters to build arrays as they develop their explanations.

(There are 3 rows of 9 or 9 columns of 3. Yes, the answer is the same both ways.)

Divide again

Show this number sentence: **9 × 3 = 27**

Can you use this number sentence to explain how to work out 27 divided by 9? (It is 3.) How can you prove your answer?

Students can use counters to build arrays as they develop their explanations.

(If you divide 27 by 3, you get 9. If you divide it by 3 again, you get 3.)

Division strategies

Show these division number sentences:

24 ÷ 4 = ______ **30 ÷ 6 =** ______

Can you tell me how to work out these problems? (4 sixes are 24. You halve 24 and then halve it again. Half of 24 is 12, half again is 12. You can take away 6 five times from 30. 30 divided by 3 is 10, divided by 3 again is 5.)

Discuss the various strategies used by students to solve the problems. Ask students to devise an appropriate word problem for one of the number sentences and explain how to solve it using words and pictures.

Double up

Materials

- counters
- number strips
- cardboard

Arrange counters in four rows of 8. Cover three rows with a piece of cardboard, showing a single row of 8 counters.

How many counters in this row? If I doubled this row, how many counters would there be? If I doubled these two rows, how many rows would there be? How many counters is that? (One row is 8. . . double of 8 is 16 . . . double 16 is 32.)

Continue the doubling pattern for 8: 8, 16, 32, 64, 128 . . . . Discuss how the result for 4 × 8 is double 2 × 8.

Halving strategy

Show this statement: 16 divided by 2 equals 8.

Can you use this statement to work out 16 divided by 4? (You halve it and then you halve it again. If you divide 16 by 2 you get 8. If you divide it by 2 again you get 4.)

How can you prove your answer?

Show a 4 × 4 array and discuss the students' suggestions.

Find what is similar

Show the multiplication tables for 2 and 4 side by side:

1 × 2 = 2	1 × 4 = 4
2 × 2 = 4	2 × 4 = 8
3 × 2 = 6	3 × 4 = 12
4 × 2 = 8	4 × 4 = 16
5 × 2 = 10	5 × 4 = 20
6 × 2 = 12	6 × 4 = 24
7 × 2 = 14	7 × 4 = 28
8 × 2 = 16	8 × 4 = 32
9 × 2 = 18	9 × 4 = 36
10 × 2 = 20	10 × 4 = 40

How are the tables similar? (The facts for 4 are double the facts for 2. The facts for 2 are half of the facts for 4.)

Variation

Repeat the activity for the tables of 3 and 6 and ask students to predict the relationship between the tables of 4 and 8.

Fast Bingo (1)

Materials

- 5 × 4 grids
- counters

What are the first 10 multiples of 4? (4, 8, 12, 16, 20, 24, 28, 32, 36, 40)

4			12
	20	8	
32		36	16
	28		
24		40	

Ask the students to randomly copy the first 10 multiples of 4 onto their grids. Make sure there is at least one number in each row and column. Ten squares will have numbers in them and 10 will be blank. This game provides quick results and maintains high interest levels. There can be many winners in each session.

Rules for Fast Bingo

1. Select a single row or column, for example, top row or third column.
2. Call out multiplication facts at random, for example, seven groups of 4.
3. Students call out the answer (28) and if they have that number in the nominated row or column, cover it with a counter.
4. When a student covers all the numbers in the row or column, they can call out 'Bingo!' to win.

Variation

Play the game using two rows or columns.

Reverse Bingo (1)

Materials

- 5 × 4 grids on paper
- counters

What are the 5 times table facts?

Ask students to randomly write the first ten 5 times table facts onto their grids. Make sure there is at least one fact in each row and column. Ten squares will have facts in them and 10 will be blank.

7 × 5			4 × 5
	1 × 5	8 × 5	
3 × 5			5 × 5
10 × 5		2 × 5	
	6 × 5		9 × 5

Rules for Reverse Bingo

1. Select a single row or column, for example, top row or third column.
2. Call out answer to the 5 times table facts at random, for example, 25.
3. Students work out the table fact (5 × 5) and if they have that fact in the nominated row or column, cover it with a counter.
4. When a student covers all the facts in the row or column, they can call out 'Bingo!' to win.

Variation

Play the game using two rows or columns, or another set of table facts.

Covering blocks

Materials

- small cardboard squares
- tens strips
- 10 × 10 array of dots

Show a 10 × 10 array and cover a block of any 25 counters, inside the array:

How many counters do you think I have covered? How many counters are not covered? (That is 5 by 4 rows . . . no, 5 by 5 . . . 25.)

Repeat the activity, covering different amounts of counters each time.

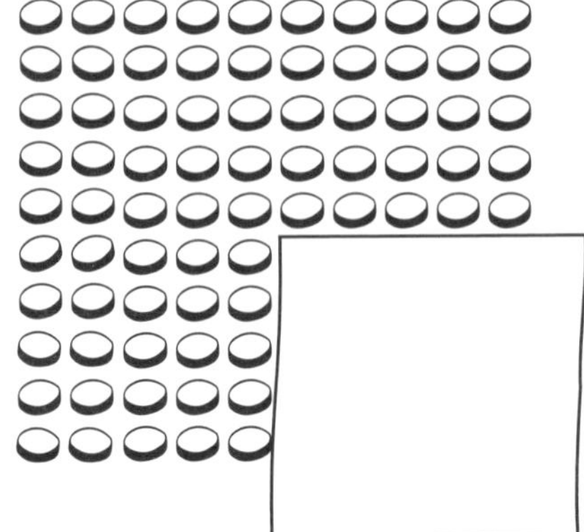

Count the eggs

Materials

- counters

Ella and Marvi were collecting eggs on the farm. They each had a small basket, with six eggs in it. How many eggs did they have altogether?

Encourage students to close their eyes and visualise the total number of eggs.

Can you use the counters to show me how many eggs there were?

Call for volunteers to arrange two groups of six counters to check the answers. Arranging the counters in arrays is not necessary at this stage, although if it arises you can take advantage of it in a discussion.

Count the pies

At a party, I saw 5 plates on the table. Each plate had 3 party pies on it. How many party pies did I see altogether? Can you imagine what the plates of pies looked like?

Encourage students to close their eyes and visualise the groups.

Can you tell me how many party pies there are altogether? (3, 6, 9, 12 . . . that is 15)

Lolly packs

Deena's Mum made up some lolly bags. She had 24 peppermints and wanted to put 6 in each bag. Can you imagine the peppermints and the number of bags?

Encourage students to close their eyes and visualise the groups.

How many bags did she need? (6, 12, 18, 24 . . . that is 4 bags.)

Sausage rolls

Ren's father bought a pack of 18 sausage rolls for the party. He put equal numbers of them on plates. Can you imagine the plates with the sausage rolls?

Encourage students to close their eyes and visualise the groups.

How many plates were there? (I think there are 3 plates. I think there are 9 plates.) How many sausage rolls were on each plate? (That is 6 rolls on each plate. There are 2 rolls on each plate.)

Sharing balloons

At the end of Erin's party, eight children shared some balloons equally. How many balloons were there? (There were 16 balloons. There were 24 balloons.) How many balloons did they each receive? (There were 2 each. There were 3 each.)

Encourage the students to represent the problem with concrete materials or diagrams. Ask students to demonstrate and discuss their answers.

Share the eggs

Last Easter, Bill and Ben got 12 eggs. If they shared the eggs between them, how many would each boy get?

Discuss the students' suggestions.

If the eggs were shared between 3 people, how many would each person get? If the eggs were shared between 4 people, how many would each person get?

Discuss the students' solutions and write the number sentence for each one.

Variation

Practise similar sharing problems, using arrays to simplify the students' solutions.

Egg groups

Show an array of 18 eggs.

If the children received 3 eggs each, how many children would there be?

Call for answers and allow students to justify their solutions. Show the number sentence for the problem:

18 ÷ 3 = ______

If the children received 6 eggs each, how many children would there be? If the children received 9 eggs each, how many children would there be?

Discuss the students' solutions and write the number sentence for each one.

Two-digit halves

Show this number: **22**

What is half of 22? (Half of 20 is 10, half of 2 is 1, 10 and 1 is 11, 11 and 11 is 22.)

Discuss the students' strategies.

Repeat the activity for: 24, 26, 28.

Variation

Extend the activity by halving numbers in this ranges: 42 to 48, 62 to 68, 82 to 88.

Year 4

Animals in your head

Discuss the number of legs on different animals, such as a bird, a dog, and an ant.

Can you work out the number of legs on 12 birds? The number of legs on 8 dogs? The number of legs on 7 ants?

Discuss the answers to the problems. Allow students to create some similar problems of their own, using different animals or different features such as wings, beaks and ears.

Number shells

Materials

- 100 counters

Arrange counters in a 6 × 6 array.

How many counters do you see altogether? (It is 6 rows of 6, that is 36.) What type of number is 36? (It makes a square, it is a square number.) Can you tell me some grouping and sharing facts for 36? (6 sixes are 36, 36 divided by 6 is 6, 36 divided by 2 is 18)

Add another shell of counters around the 6 × 6 array:

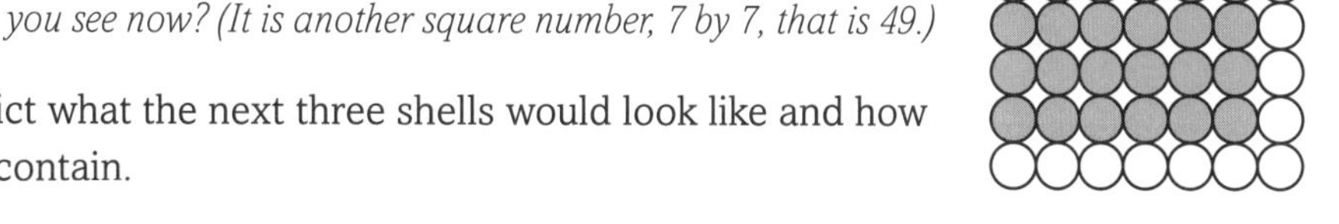

How many counters do you see now? (It is another square number, 7 by 7, that is 49.)

Ask students to predict what the next three shells would look like and how many numbers they contain.

Making arrays

Materials

- 60 counters

Show a collection of 60 counters.

Can you estimate how many counters there are here?

Discuss how students developed their estimates before giving the answer.

How many different arrays could we make using 60 counters? How many counters long and wide will each array be? (10 by 6, I have got 4 fifteens, 12 groups of 5 make 60, 3 rows of 20, 2 columns of 30, one row of 60)

Encourage students to work in pairs to discover arrays of different sizes.

Factors

Ask students to recall all the different multiplication facts for 60 (as above).

Does anyone know the word we use to describe numbers that multiply together to make other numbers?

Introduce the term 'factor' to describe numbers which multiply together to make other numbers. Repeat the activity, listing factors for 72 and 96.

Variation

Repeat the activity for other 2-digit multiples.

Matter of facts

Materials

- 48 counters

Show counters in an 8 × 6 array.

How many counters do you see altogether? (It is 8 rows of 6, that is 48.) Can you tell me some grouping and sharing facts for 48? (8 sixes are 48, 48 divided by 2 is 24, one-quarter of 48 is 12 . . .)

Variation

Repeat the activity often for other arrays.

Mixed fact rectangles

Materials

- 26 red counters
- 24 yellow counters

Show the counters in a 10 × 5 array with the red counters in a shell around the yellow counters:

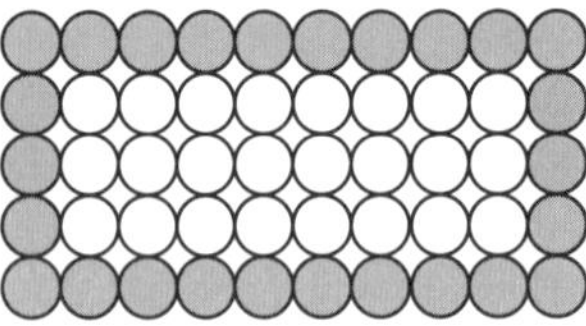

How many counters do you see altogether? How did you get that answer? (It is 2 rows of 10 plus 2 columns of 3, 26. It is 2 fives plus 2 eights. I multiplied 10 by 5, that is 50.)

Discuss the students' responses.

How could you find out how many red counters are there? (You can multiply 10 by 5 and take away the inside rectangle which is 8 × 3, that is 50 take away 24, that is 26.)

Discuss the students' responses and record them as number sentences:

2 × 10 + 2 × 3 = 26 **2 × 5 + 2 × 8 = 26** **10 × 5 – 8 × 3 = 26**

Mixed fact squares

Materials

- 20 red counters
- 16 yellow counters

Show counters in a 6 × 6 array with the red counters in a shell around the yellow counters:

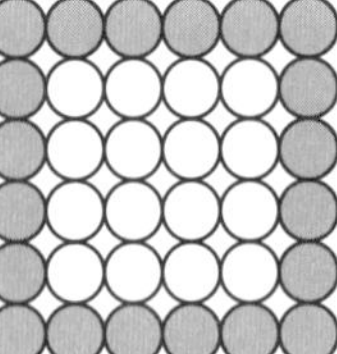

How could you find out how many red counters are there? (It is 2 rows of 6 plus 2 columns of 4, 20. You can count 5 on each side, 4 fives are 20. It is 4 fours on each side plus the 4 corners. It is 4 sixes on each side take away the 4 corners, that is 20.)

Record the students' responses as number sentences:

2 × 6 + 2 × 4 = 20 **4 × 5 = 20** **4 × 4 + 4 = 20** **4 × 6 – 4 = 20.**

Variation

Repeat the activity for 64, 81 and 100.

Multiples

Materials

- 63 counters

Show counters in a 9 × 7 array.

How many counters do you see? What are the factors of 63? (The factors of 63 are 9 and 7.)

Remind students that the product of the factors of a number is called a 'multiple'.

What is a multiple of 9 and 7? (63)

Complete these statements with the students:

_______ is a multiple of 6 and 5.

_______ is a multiple of 5 and 9.

_______ is a multiple of 9 and 4.

_______ is a multiple of 8 and 4.

_______ is a multiple of 7 and 8.

_______ is a multiple of 6 and 9.

Fast Bingo (2)

Materials

- 5 × 4 grids
- counters

Can you tell me the first 10 multiples of 7? (7, 14, 21, 28, 35, 42, 49, 56, 63, 70)

Ask the students to randomly copy the first 10 multiples of 7 onto their grids. Make sure there is at least one number in each row and column. Ten squares will have numbers in them and 10 will be blank. This game provides quick results and maintains high interest levels. There can be many winners in each session.

7			21
	35	14	
56		63	28
	49		
42		70	

Rules for Fast Bingo

1. Select a single row or column, for example, top row or third column.
2. Call out multiplication facts at random, for example, six groups of 7.
3. Students call out the answer (42) and if they have that number in the nominated row or column, cover it with a counter.
4. When a student covers all the numbers in the row or column, they can call out 'Bingo!' to win.

Variation

Play the game using two rows or columns.

Reverse Bingo (2)

Materials

- 5 × 4 grids on paper
- counters

What are the 9 times table facts?

Ask students to randomly write the first ten 9 times table facts onto their grids. Make sure there is at least one fact in each row and column. Ten squares will have facts in them and 10 will be blank.

7 × 9			4 × 9
	1 × 9	8 × 9	
3 × 9			5 × 9
10 × 9		2 × 9	
	6 × 9		9 × 9

Rules for Reverse Bingo

1. Select a single row or column, for example, top row or third column.
2. Call out answer to the 9 times table facts at random, for example, 27.
3. Students work out the table fact (3 × 9) and if they have that fact in the nominated row or column, cover it with a counter.
4. When a student covers all the facts in the row or column, they can call out 'Bingo!' to win.

Variation

Play the game using two rows or columns, or another set of table facts.

Fact reckoner

Materials

- multiplication grids
- coloured pencils

Give each student a multiplication grid.

Which table facts have you memorised? Which facts do you still need to know?

Ask the students to colour in the table facts that they have memorised. The grid can then be glued in maths books. More table facts can be coloured in later as they are learnt progressively. Remind students that because of the commutative law, the number of facts to be learned is approximately halved.

×	1	2	3	4	5	6	7	8	9	10
1	1	2	3	4	5	6	7	8	9	10
2	2	4	6	8	10	12	14	16	18	20
3	3	6	9	12	15	18	21	24	27	30
4	4	8	12	16	20	24	28	32	36	40
5	5	10	15	20	25	30	35	40	45	50
6	6	12	18	24	30	36	42	48	54	60
7	7	14	21	28	35	42	49	56	63	70
8	8	16	24	32	40	48	56	64	72	80
9	9	18	27	36	45	54	63	72	81	90
10	10	20	30	40	50	60	70	80	90	10

How many legs?

At the nature reserve, I saw some wombats and emus. I counted 40 legs all together. How many wombats were there? How many emus?

Show students a picture of a wombat and an emu. Have students work in pairs and present their strategies to the class. They may use diagrams or concrete materials in their explanations.

Variation

Solve similar problems involving 36 legs, 60 legs and 96 legs.

Larger doubling

Show this number: **86**

What is this number doubled?

Discuss the students' strategies.

(You double the tens and double the units. Add them together. That is 16 tens and 12 ones, 172. You can double the 90 and take away 8.)

Variation

Extend the activity for numbers from 50 to 99.

Larger halving

Show this number: **56**

What is this number halved?

Discuss the students' strategies.

(You halve the tens and halve the units. Add them together. That is 25 and 3, 28.)

Variation

Extend the activity for even numbers from 32 to 98.

Digit problem

How many toes are there altogether in our class today? Can you tell me how you worked it out?

Discuss the strategy of multiplying by 10.

Darren and Toula delivered 900 catalogues in 10 days. How many did they deliver on average each day?

Discuss the strategy of dividing by 10.

Multiply by 10

Show these number sentences: **6 × 10**, **16 × 10**, **60 × 10**

Can you tell me how to work out these problems? Why do we add a zero when we multiply by 10?

Discuss the students' responses. With the aid of the place value chart, explain that multiplying by 10 shifts everything one place to the left. The number becomes larger by the power of 10 and a zero is inserted as a place holder.

Hundreds	Tens	Ones

Divide by 10

Show these number sentences: **80 ÷ 10**, **180 ÷ 10**, **280 ÷ 10**

Can you tell me how to work out these problems? Why do we cross out a zero when we divide by 10?

Discuss the students' responses. With the aid of the place value chart, explain that dividing by 10 shifts everything one place to the right. The number becomes smaller by a power of 10 and the zero is therefore deleted.

Hundreds	Tens	Ones

Five strategy

Show this number sentence: **5 × 18**

Can you tell me how to work out this problem? (You multiply the tens and then the ones, 5 tens is 50, 5 eights are 40, 50 and 40 equals 90. There is an easier way, multiply by 10 and then halve it, 10 eighteens is 180, half of that is 90. 5 × 18 is double 5 × 9. That is double 45, so it is 90.)

Discuss and record the various strategies used by the students.

Variation

Extend to other decades as appropriate, for example, 5 × 22, 5 × 44, 5 × 68, 5 × 86.

Multiply by 20

Show this multiplication number sentence: **8 × 20**

Can you tell me how to work out this problem? (I think it is 8 groups of 2 tens, that is 160. It is the same as 8 by 2 by 10, so 8 twos are 16 and 16 multiplied by 10 is 160.)

Discuss and record the strategies used by students. Ask the students to devise an appropriate word problem for the number sentence.

Multiply by 40

Show this multiplication number sentence: **7 × 40**

Can you tell me how to work out this problem? (I multiply 4 tens by 7, that is 28 tens or 280. To multiply by 40, you double the number, double it again and multiply by 10 . . . 7, 14, 28, 280.)

Discuss and record the various strategies used by students.

Extension

Repeat the activity using these number sentences:

6 × 40 **8 × 40** **9 × 40** **10 × 40** **11 × 40** **12 × 40**

Multiply by 100

Materials

- a place value chart

Show these number sentences:

7 × 100 = ______ **17 ×100 =** ______ **72 × 100 =** ______

Can you tell me how to work out these problems? (It is easy to multiply by 100. You just add two zeros.) Why do we add two zeros when we multiply by 100?

Th	H	T	U	Total
		1	7	
1	7	0	0	

Discuss the students' responses. With the aid of the place value chart, explain that multiplying by 100 shifts everything two places to the left. The number becomes larger by two powers of 10 and two zeros are inserted as place holders.

Divide by 100

Materials

- a place value chart

Show these number sentences:

80 ÷ 100 = ______ **1800 ÷ 100 =** ______ **2800 ÷ 100 =**______

Can you tell me how to work out these problems? (It is easy to divide by 100. You just cross out 2 zeros.) Why do we cross out two zeros when we divide by 100?

Discuss the students' responses. With the aid of the place value chart, explain that dividing by 100 shifts everything two places to the right. The number becomes smaller by two powers of 10 and the two zeros are therefore deleted.

÷ 100 → 2 places

Th	H	T	U	Total
1	8	0	0	
		1	8	

Multiply by 50

Show this multiplication number sentence: **8 × 50**

Can you tell me how to work out this problem? (I multiply 5 tens by 8, that is 40 tens or 400.

To multiply by 50, you can multiply by 100 and then halve it, that is 800 halved . . . 400.)

Discuss and record the various strategies used by the students.

Extension

Repeat the activity with these number sentences:

6 × 50 **7 × 50** **9 × 50** **8 × 50** **11 × 50** **12 × 50**

Divide by 20

Show this division number sentence: **140 ÷ 20 =** ______

Can you tell me how to work out this problem? (That is 140 divided by 2 and then divided by 10. It is 7. I think of multiplication. 7 twos are 14 so 7 groups of 2 tens are 14 tens. The answer is 7.)

Discuss and record the various strategies used by the students.

Extension

Repeat the activity with these number sentences:

160 ÷ 20 **180 ÷ 20** **220 ÷ 20** **240 ÷ 20**

Divide by 40

Show this division number sentence: **240 ÷ 40 =** ______

Can you tell me how to work out this problem? (To divide by 40 you just halve 240, halve it again and divide by ten, that's 120 . . . 60 . . . divided by 10 is 6. Six fours are 24 so 6 times 40 is 240.)

Discuss and record the various strategies used by the students. Ask the students to devise an appropriate word problem for the number sentence.

Divide by 50

Show this division number sentence: **400 ÷ 50 =** ______

Can you tell me how to work out this problem?

Discuss and record the various strategies used by the students. (8 fives equal 40, 8 times 5 tens makes 40 tens . . . that is 400. The answer is 8. I divide by 100 and then double it, that is 4 doubled makes 8.)

Extension

Repeat the activity with these number sentences:

200 ÷ 50 **500 ÷ 50** **600 ÷ 50** **800 ÷ 50**

Halving and halving

Nick's father shared 1 L of orange juice equally among 4 people. How much juice did each person receive? (You can do it by halving. Half of 1 L is 500 mL. Halve it again and you get 250 mL.)

Ask pairs of students to present their strategies to the class.

Variation

Ask the students how much juice each person would receive if it was divided among 8 people.

Timber cuts

How many lengths of timber would there be if a 2-m length of timber was cut into 100-cm lengths . . . 50-cm lengths . . . 40-cm lengths . . . 20-cm lengths?

Ask the students to work in pairs and present their strategies to the class. Students may use diagrams or concrete materials in their explanations. Relate the discussion to previous strategies developed to divide by 100, 50, 40 and 20.

FRACTIONS AND DECIMALS

Year 3

Fraction meanings

Show students pictures of a glass of water, an aerial view of a netball court, a 1-L jug, a metre ruler – or use real-life objects if available. Ask students to draw or indicate $\frac{3}{4}$ of the glass of water, $\frac{2}{3}$ of the area of the netball court, $\frac{1}{8}$ of the 1-L jug and $\frac{1}{10}$ of the metre ruler. Discuss the students' representations and suggestions.

Extension

Discuss the remaining fraction of each object that is not represented.

Comparing cars

Show a group of four cars with three cars of one colour and one car of another colour, for example red.

What fraction of the cars is red? $\left(\frac{1}{4}\right)$

Discuss the students' responses.

What fraction of the cars is not red? $\left(\frac{3}{4}\right)$

Discuss the students' responses.

Which fraction is larger, $\frac{1}{4}$ *or* $\frac{3}{4}$*? How do you know? (Three-quarters is larger because you have three equal parts of four.)*

Discuss the students' answers in terms of the relative sizes of the fractions. Ensure that students understand that as the denominator (the number of equal parts the whole has been divided into) is the same, the larger fraction is the one with the larger numerator (number of equal parts).

Jellybean fractions

Show a group of eight jellybeans, including three that are yellow.

What fraction of the jellybeans is yellow? $\left(\frac{3}{8}\right)$

Discuss the students' responses.

What fraction of the jellybeans is not yellow? $\left(\frac{5}{8}\right)$

Discuss the students' responses.

Which fraction is larger, $\frac{3}{8}$ *or* $\frac{5}{8}$*? How do you know? (Five-eighths is larger because you have 5 equal parts of 8.)*

Discuss the students' answers in terms of the relative sizes of the fractions.

Pizza halves

If I have a pizza made up of 8 equal slices, and I share this pizza equally between 2 people, how many slices will they each get?

If necessary, show students a picture of the pizza.

What fraction of the pizza would each person get? $\left(\frac{1}{2}\right)$

Ask the students to explain their answers.

How many ways can you divide this pizza into halves?

Discuss the students' suggestions.

Pizza shares

If I have a pizza made up of 8 equal slices, and I share this pizza equally between 8 people, how many slices will they each get?

If necessary, show students a picture of the pizza.

What fraction of the pizza would each person get? $\left(\frac{1}{8}\right)$ *If I shared this pizza equally among four people, how many slices would they each get? (Two each.) What fraction of the pizza would each person get?* $\left(\frac{1}{4}\right)$ *Which fraction is larger,* $\frac{1}{4}$ *or* $\frac{1}{8}$*? How do you know?*

Discuss the students' answers in terms of the relative sizes of the fractions.

Chocolate strips

Show a ten-strip to represent a bar of chocolate containing 10 squares.

If I ate one square of chocolate, what fraction of the chocolate bar would that be? (One out of ten – one-tenth.)

Colour in or cover one square.

What fraction of the chocolate bar would be left? $\left(\frac{9}{10}\right)$

Variation

Repeat the activity, based on eating four squares of chocolate.

Half of one-half

Nick had half a chocolate bar. He divided his piece in half again to share with his brother. What fraction of the original chocolate bar did the boys have now? (Each boy had half of a half, or one-quarter of the whole bar.) How did you work out the answer?

Discuss the different strategies used by students. Highlight the strategy of using a diagram to help solve the problem.

How many each?

Three children shared a packet of 16 jellybeans. The first child received $\frac{1}{8}$ *of the jellybeans, the second child received* $\frac{1}{4}$ *and the third child received* $\frac{1}{2}$*. How many jelly beans did each child receive? (Two, four, eight) Were any left over? (Two) How would you work out this problem?*

Discuss the different strategies used by students. Encourage the students to help solve the problem using a diagram or concrete materials.

A third or a quarter?

Two children were offered fractions of a packet of jelly beans. They could take $\frac{1}{3}$ *or* $\frac{1}{4}$*. Which fraction did they take? Why? (If there were 12 jellybeans,* $\frac{1}{3}$ *is 4 and* $\frac{1}{4}$ *is 3, so I would take* $\frac{1}{3}$ *because it is bigger, but if you did not like jellybeans, you might take* $\frac{1}{4}$ *because it is smaller.)*

Discuss the different strategies used by students. Encourage the students to use diagrams or concrete materials to help solve the problem if necessary.

Taking the cake

Two children were offered a slice of birthday cake. They could take either $\frac{1}{8}$ or $\frac{1}{4}$ of the cake.

Which fraction did they take? Why? (If you cut the cake into 8 slices, you will see that $\frac{1}{4}$ is the same as $\frac{2}{8}$. If I was hungry, I would take one-quarter because it is bigger than $\frac{1}{8}$. If I was not hungry, or if there was not enough cake for everyone, I would take $\frac{1}{8}$ because it is smaller.)

Discuss the different strategies used by students. Encourage students to use diagrams or concrete materials to help solve the problem if necessary.

Variation

Repeat the activity for other fractions such as $\frac{1}{2}$ and $\frac{2}{5}$.

Year 4

Fraction bars

Show these three fraction bars:

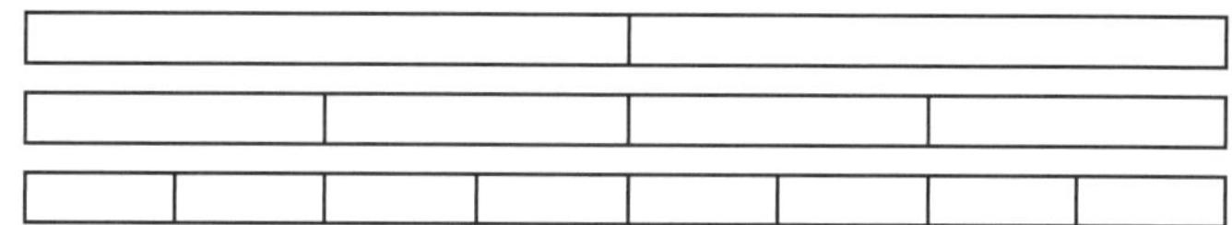

Ask students to shade one-half of the top bar, two-quarters of the middle bar and four-eighths of the bottom bar. Label the bars or show the fractions: $\frac{1}{2}$, $\frac{2}{4}$, $\frac{4}{8}$.

What do you notice about one-half, two-quarters and four-eighths? (They are the same length. They take up the same space.)

Discuss the students' responses. Introduce the term 'equivalent fractions' to talk about fractions of this type.

Equivalent fractions

Show these fractions: $\frac{1}{2}$, $\frac{2}{4}$, $\frac{4}{8}$.

Can you see a pattern or relationship between these fractions? (If you double the top and bottom of $\frac{1}{2}$, you get $\frac{1}{4}$. If you divide 4 by 2 you get 2. 8 divided by 2 is 4. That is two-quarters.)

Discuss the students' responses. Reinforce the students' understanding of equivalent fractions, using fraction bars or a fraction wall if necessary. Ask students to suggest other equivalent fractions for $\frac{1}{2}$ such as $\frac{3}{6}$, $\frac{5}{10}$, $\frac{6}{12}$, $\frac{8}{16}$.

Quarter line

Materials

- 1-m length of string
- number cards for 0 and 1
- pegs
- fraction cards for $\frac{1}{4}$, $\frac{2}{4}$, $\frac{3}{4}$ and $\frac{4}{4}$

Construct a number line segment from 0 to 1.

Choose students to randomly select a fraction card.

Can you estimate where to place this fraction card? Why did you place it there?

Attach the cards in the position suggested by the students and discuss their suggestions. Count in quarters forwards and backwards along the number line. Discuss the equivalence of $\frac{4}{4}$ and 1 with the students.

Eighths line

Materials

- 2-m length of string
- number cards for 0 and 1
- pegs
- fraction cards for $\frac{1}{8}$, $\frac{2}{8}$, $\frac{3}{8}$, $\frac{4}{8}$, $\frac{5}{8}$, $\frac{6}{8}$, $\frac{7}{8}$, $\frac{8}{8}$

Construct a number line segment from 0 to 1. Choose students to randomly select a fraction card.

Can you estimate where to place this fraction card? Why did you place it there?

Attach the cards in the position suggested by the students and discuss their suggestions. Count in eighths forwards and backwards along the number line. Discuss the equivalence of $\frac{8}{8}$ and 1 with the students.

Variation

Repeat the activity for fifths and tenths.

Mixed numerals

Show 3 squares:

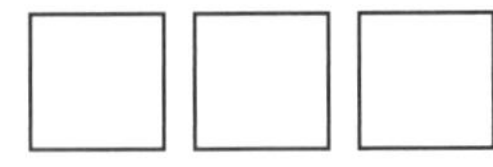

How could you show the idea of $2\frac{1}{2}$ using these 3 squares?

Use the squares to model the mixed numeral, by covering or removing half of the third square.

How many halves in $2\frac{1}{2}$?(Five) How do you know?

Discuss the students' responses. Count the number of halves in the squares with the class.

Quarter line beyond one

Materials

- 2-m length of string
- number cards for 0, 1 and 2
- pegs
- fraction cards for $\frac{1}{4}$, $\frac{2}{4}$, $\frac{3}{4}$, $1\frac{1}{4}$, $1\frac{2}{4}$, $1\frac{3}{4}$

Construct a number line segment from 0 to 1.

Choose students to randomly select a fraction card.

Can you estimate where to place this fraction card? Why did you place it there?

Attach the cards in the position suggested by the students and discuss their suggestions. Count in quarters forwards and backwards along the number line.

Variation

Repeat the activity for other fraction intervals (fifths, eighths and tenths) from 0 to 2.

Covering hundredths

Materials

- a blank hundreds chart
- thin cardboard strips

Show a blank hundreds chart. Tell students that the squares represent a large block of chocolate which can be divided. Emphasise the idea of division of one whole. Cover 46 squares.

What fraction of the block is covered? (46 squares, 46 hundredths, that is $\frac{46}{100}$ or 0.46) What fraction of the block is not covered? (54 squares, 54 hundredths, that is $\frac{54}{100}$ or 0.54)

Three-way notation

Materials

- a blank hundreds chart
- thin cardboard strips

Show a blank hundreds chart. Tell students that the squares represent a large block of chocolate that can be divided. Emphasise the idea of division of one whole. Invite a student to cover any number of squares on the block with cardboard strips.

What fraction of the squares is covered? How many ways can we write the fraction that was covered?

Discuss the students' responses. Model recording of the amount as a fraction, decimal and a percentage, so for 37 squares or 37 hundredths: $\frac{37}{100}$, 0.37, 37%

Repeat the activity for the fraction of the block that was not covered.

Equivalent decimals

Materials

- a blank hundreds chart
- thin cardboard strips

Cover a row or column of 10 squares on the hundreds chart with a cardboard strip.

What decimal fraction has been covered? (10 hundredths) Does anyone else know another name for the same decimal fraction? (1 tenth) Can anyone explain why they are the same?

Discuss the idea of equivalence with the students. Repeat the activity with other equivalent decimal fractions, for example, 30 hundredths and 3 tenths.

Decimal order

Materials

- a set of number cards from 0 to 9
- a decimal place value chart

Ask three pairs of students to randomly choose a number card. Each pair uses the cards to make a decimal number less than one on the place value chart, for example, 0.43, 0.78, 0.69.

Hund	Tens	Ones •	Tenths	Hundredths

Which of these decimals is largest? Which is smallest? Can you place the decimals in order from largest to smallest? How do you know which decimal is largest or smallest?

Discuss the students' responses in terms of place value. List the decimals in order from largest to smallest. Repeat the activity with different pairs of students.

Numeral card subtraction

Materials

- a set of number cards from 0 to 9
- a decimal place value chart

Ask two pairs of students to randomly choose a card. Each pair of students uses the cards to make a decimal number less than one on the place value chart, for example, 0.34, 0.69.

Hund	Tens	Ones •	Tenths	Hundredths

What is the answer if you subtract the smaller from the larger of these two decimal numbers? (0.35)

Discuss the different strategies used by students to solve the problem.

Variation

The question may also be posed as finding the difference between the two decimals.

Numeral card addition

Materials

- a set of number cards from 0 to 9
- a decimal place value chart

Hund	Tens	Ones •	Tenths	Hundredths

Ask two pairs of students to randomly choose a card. Each pair of students uses the cards to make a decimal number less than one on the place value chart, for example, 0.24, 0.63.

What is the answer if you add the two decimal numbers? (0.87)

Discuss the different strategies used by students to solve the problem.

Decimal line up

Materials

- 1-m length of string
- pegs
- number cards for 0 and 1
- decimal cards for 0.1, 0.2, 0.3, 0.4, 0.5, 0.6, 0.7, 0.8 and 0.9

Construct a number line segment from 0 to 1.

Choose students to randomly select a decimal card.

Can you estimate where to place this decimal card? Why did you place it there?

Attach the cards in the position suggested by the students and discuss their suggestions. Count the decimals forwards and backwards along the number line.

Variation

Repeat the activity with percentage cards.

Mixed decimals

Materials

- 2-m length of string
- pegs
- number cards for 0 and 1
- 10 to 15 decimal cards to two decimal places, for example, 0.23, 0.35, 0.47

Construct a number line segment from 0 to 1.

Choose students to randomly select a decimal card.

Can you estimate where to place this decimal card? Why did you place it there?

Attach the cards in the position suggested by the students and discuss their suggestions. Count the decimals forwards and backwards along the number line.

Extension

Repeat the activity with mixed cards, showing fractions, decimals and percentages.

Long jump

Materials

- four 1-m rulers

Show a number line segment between 3 and 4, with 10 intervals:

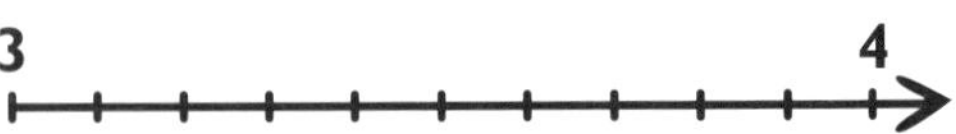

If Katie jumped 3.73 m in the long jump competition, where would that be on the number line? How many centimetres in 3.73 m? (373) How do you know?

Discuss the students' suggestions. Relate the discussion to a number line beyond 1 with centimetre divisions. Demonstrate the distance using four 1-m rulers. Ask students to convert these measurements to metres, using decimals:

135 cm **327 cm** **631 cm** **265 cm**

Decimals beyond one

Materials

- 2-m length of string
- pegs
- number cards for 0, 1 and 2
- a set of decimal cards in intervals of 0.1 from 0 to 2

Construct a number line segment from 0 to 2.

Choose students to randomly select a decimal card.

Can you estimate where to place this decimal card? Why did you place it there?

Attach the cards in the position suggested by the students and discuss their suggestions. Count the decimals forwards and backwards along the number line.

Variation

Repeat the activity for other decimal intervals (0.2, 0.25) from 0 to 2.

MONEY AND FINANCIAL MATHEMATICS

Year 3

Coin combos

I have a 20c coin, a $1 coin, four 10c coins, two 50c coins and three 5c coins. How can I use these coins to buy a $1.20 icy pole? How much money do I have left over? (The easiest way is to give the dollar and the 20-c coin. If you wanted to get rid of your change, you could give 2 fifties, 2 fives and 1 ten.)

Discuss the students' suggestions, listing amounts and totals.

Variation

Students can devise similar coin combo problems for a partner to solve.

Extension

Ask students to suggest the least number of coins needed, the greatest number of coins needed, or the least number of silver coins needed.

Compare it

Alex and Yasmine were saving to buy bicycles. Alex saved $122 and Yasmine saved $86. How much more money did Alex have than Yasmine?

What is the best method to work out this problem?

Discuss the different strategies used by students. Some students may wish to use a number line to count on, while others may use the written algorithm or concrete materials.

How much more?

Materials

- a class set of calculators

Kim has saved $67 towards a game that costs between $180 and $200. About how much more does Kim need to save? ($130, $120)

Invite the students to work in pairs and discuss their estimations. Discuss the various strategies used. Ask students to complete the problem by setting an amount for the purchase and calculating the answer. Calculators may be used.

The school shop

Show this price list and ask students to work in pairs to solve these problems mentally.

Ted bought the two most expensive items on the list. How much did he pay? Ruby bought the two cheapest items. How much did she pay? How much more do coloured pencils cost than a notepad? If you bought felt pens and coloured pencils, how much change would you receive from $10.00? Shukri spent exactly $5.00. What did he buy?

SCHOOL SUPPLIES:	
Exercise books	$1.60
Rulers	$1.10
Coloured pencils	$3.00
Pens	$2.40
Notepads	$0.90
Erasers	$1.00

Extension

Students can devise their own problems from the list and ask classmates to solve them.

Year 4

Take away

Naja had $537 in her bank account. She withdrew $127. How much money was left in her account? (You take away 100 first, that is 437 . . . take away 27 more which leaves 410.) What strategy would you use to solve this problem?

Discuss the different strategies used by the students.

Enough money?

Jack wants to buy a bike that costs about $700. He has $345 in a savings account and $371 in a jar. Does Jack have enough money to buy the bike? How do you know?

Invite students to work in pairs to discuss their estimation strategies and present their thoughts to the rest of the class. Discuss the various strategies used. Ask the students to complete the problem by setting an amount for the purchase and calculating the answer.

Birthday games

Thomas received $100 for his birthday. He went to the games shop and saw games for $24.95, $19.95, $49.50 and $34.50. Which games could he buy? (He cannot buy them all because the total is about $130.)

Discuss the students' strategies. If necessary, discuss the advantages of estimation instead of direct calculation.

Variation

Discuss situations in the students' lives where estimations, rather than direct calculations, are most useful.

Canteen lunch

Sandwiches $3.50, Rolls $4.25, wraps $4.80, sushi $3.20, fruit salad $4.50, fruit $1.20

Andre purchased two sushi rolls and a piece of fruit. How much change did he receive from $10.00? How many wraps can you buy for $20.00? Would there be any change?

What would you purchase from the canteen if you had $10.00? Would there be any change?

CANTEEN LUNCH:	
Sandwiches	$3.50
Rolls	$4.25
Wraps	$4.80
Sushi	$3.20
Fruit salad	$4.50
Fruit	$1.20

What would you purchase if you had $20.00? Would there be any change?

Allow students time to create a lunch order, calculate the cost and predict the change. Ask students to share their lunch orders with the class.

PATTERNS AND ALGEBRA

Year 3

Chart count

Materials

- a hundreds chart

Show the hundreds chart and place a mark or counter on 7:

Can you count in tens down the chart? (7, 17, 27 . . .97)

Cover or remove the hundreds chart and ask students to repeat the activity. This will help students' mental imagery of the patterning of the numbers.

Can you imagine the numbers on the chart as you count them?

1	2	3	4	5	6	7	8	9	10
11	12	13	14	15	16	17	18	19	20
21	22	23	24	25	26	27	28	29	30
31	32	33	34	35	36	37	39	40	41
41	42	43	44	45	46	47	48	49	50
51	52	53	54	55	56	57	58	59	60
61	62	63	64	65	66	67	68	69	70
71	72	73	74	75	76	77	78	79	80
81	82	83	84	85	86	87	88	89	90
91	92	93	94	95	96	97	98	99	100

Repeat the activity for the number 92, this time counting backwards in tens, with and without the chart.

Variation

Count forwards and backwards by tens from a variety of different numbers.

Line count (1)

Show a number line segment from 0 to 200, with intervals of 10. This is one section:

Can you estimate where 190 would be?

Discuss the students' responses and mark the position on the number line.

Can you count backwards by tens from 190? (190, 180, 170 . . . 0)

Can you count forwards by tens from 0? (10, 20, 30 . . . 200)

Encourage the whole class to count forwards by tens.

Variation

Count forwards and backwards by tens from a variety of numbers on the decade. Use different number lines segments such as 200 to 400, or 500 to 700.

Line count (2)

Show a number line segment from 0 to 200, with intervals of 10. This is one section:

Can you estimate where 176 would be?

Discuss the students' responses and mark the position on the number line.

Can you count backwards by tens from 176? (176, 166, 156 . . . 6)

Can you count forwards by tens from 6? (6, 16, 26 . . .196)

Variation

Count forwards and backwards by tens from a variety of numbers off the decade. Use different number lines segments such as 200 to 400, or 500 to 700.

Line count (3)

Show an open number line segment from 0 to 500:

Can you estimate where 440 would be?

Discuss the students' responses and mark the position on the number line.

Can you count backwards by hundreds from 440? (440, 340, 240, 140, 40) Can you count forwards by hundreds from 40? (40, 140, 240 . . .440)

Variation

Count forwards and backwards by hundreds from a variety of numbers on the decade. Use a number line segment from 500 to 1000.

Line count (4)

Show an open number line segment from 0 to 1000:

0 ——————————— 1000

Can you estimate where 970 would be?

Discuss the students' responses and mark the position on the number line.

Can you count backwards by hundreds from 970? (970, 870, 770 . . . 70) Can you count forwards by hundreds from 70? (70, 170, 270 . . . 970)

Encourage whole class counting forwards by hundreds.

Variation

Count forwards and backwards by hundreds from a variety of numbers on the decade.

Line count (5)

Show an open number line segment from 0 to 1000:

0 ——————————— 1000

Can you estimate where the number 826 would be?

Discuss the students' responses and mark the position on the number line.

Can you count backwards by hundreds from 826? (826, 726, 626 . . . 26) Can you count forwards by hundreds from 26? (26, 126, 226 . . . 926)

Variation

Count forwards and backwards by hundreds from a variety of numbers off the decade.

Stick triangles

Materials

- craft sticks

Show this pattern of triangles made with craft sticks:

Can you tell me what the pattern is? (It is adding two sides each time. You keep on making triangles.)

Discuss the students' responses and continue demonstrating the pattern.

Can you tell me the number of sticks in each triangle?

Show the numbers 3, 5, 7, 9, 11, 13 underneath the first six triangles.

Can anyone tell me what these types of numbers are called? (They are the odd numbers. You cannot share them evenly into two groups.)

Stick squares

Show this pattern of squares made with craft sticks:

Can you tell me what the pattern is? (It is adding three sides each time. You keep on making squares.)

Discuss the students' responses and continue demonstrating the pattern.

Can you tell me the number of sticks in each square?

Show the numbers 4, 7, 10, 13, 16, 19 underneath the first six squares.

Can anyone predict the next three numbers?

Discuss the strategies used by the students.

Calculator patterns

Materials

- a class set of calculators
- pen and paper

Do you know how to count on using the calculator?

Demonstrate the use of the constant function. Enter [5] [+] [=] and then continue to press [=] so that the pattern of fives appears. Ask a student to suggest a 2-digit number, for example, 45.

Do you know how to count on by sixes from 45?

Enter [4] [5] [+] [6] [=] and then continue to press [+] so that the calculator generates a pattern which counts by sixes. Ask the students to enter a 2-digit number on their calculator and investigate counting forwards by fives, sixes, sevens or nines.

Introduce the recording of answers. Students can start at any 2-digit number, press [+] and a number of their choice, press [I] and record the number each time. Limit the activity to 10 key presses. Note that on some calculators, the symbol for 1 may need to be entered twice.

Backwards patterns

Materials

- a class set of calculators
- pen and paper

Do you know how to count backwards using the calculator?

Demonstrate the procedure. Enter [I] [0] [0] [–] [5] then press [+].

Continue to press 5 [5] so that 5 [5] is subtracted each [5] time. Ask the students to start with a 3-digit number on their calculator and investigate counting backwards by fives, sixes, sevens, eights or nines.

Introduce the recording of answers. Students can start at any 3-digit number, press [–] and a number of their choice, press [=] and record the number each time. Limit the activity to 10 key presses. Note that negative numbers will arise in this activity and may be further discussed with the students by using a number line.

Three and four patterns

Show these frames with the numbers:

18	21	24			
40	36	32			
158	162	166			
237	234	231			

Can you see a pattern in each frame? (It is going up by 3. It is going up by 4. It is going down by 4.)

Complete each frame as the students supply the answers. Ask the class to count the numbers in each pattern, forwards and backwards.

Missing patterns

Show these frames with the numbers:

141		121	111		
	654		658	660	
256		262		268	
555		545		535	

Can you see a pattern in each frame? (You take away 5 . . . . It is adding on 10 . . .)

Complete each frame as the students supply the answers. Ask the class to count the numbers in each pattern, forwards and backwards.

Year 4

Counting chart patterns

Materials

- a blank hundreds chart

Show the chart and mark this pattern:

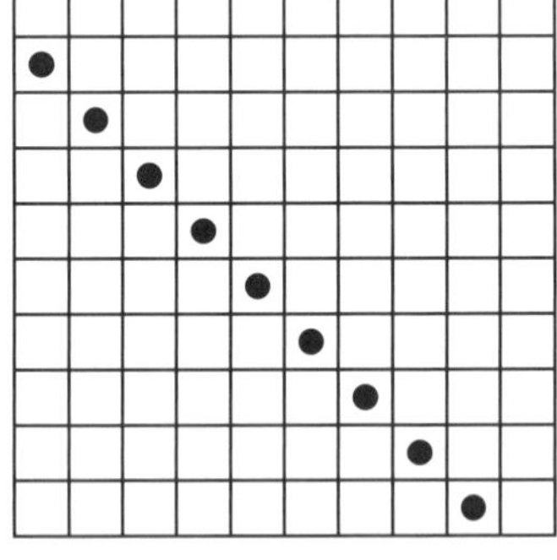

What can you tell me about the pattern? (It is diagonal. It is the 11 times table facts.)

Show the multiples of 11 on the grid and discuss the 11 times table.

Can you see anything unusual about the pattern of numbers in the 11 times table? (The tens and ones digits are the same. As you count, they both go up by 1.)

Discuss the students' responses. Count by 11 times forwards and backwards.

Nines pattern

Materials

- a blank hundreds chart

Show the chart and mark this pattern:

What can you tell me about the pattern? (It is diagonal. It is the table of nines.)

Show the multiples of 9 on the grid and discuss the table facts for nines.

Can you see anything unusual about the pattern of numbers in the table facts for nines? (As you count, the tens go up by 1 and the ones go down by 1. If you add the 2 digits of a number in the nines pattern, they always add up to 9.)

Discuss the students' responses. Count by nines forwards and backwards.

Variation

Investigate the pattern of sevens and eights with the students.

Fours and eights

Materials

- a blank hundreds chart

Show the chart and mark this pattern:

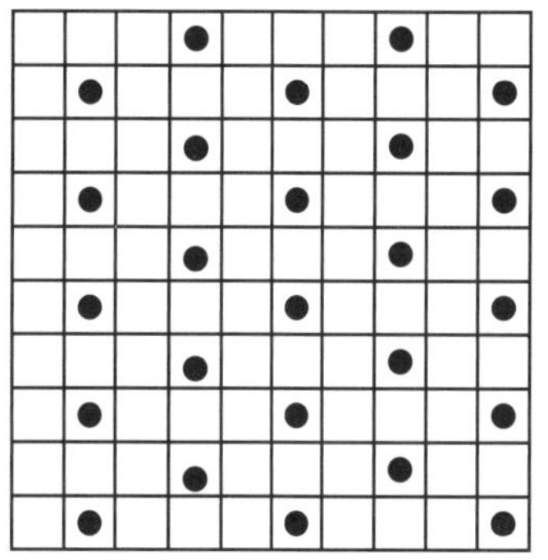

What can you tell me about the pattern? (It is going up by 4 each time. It is a counting by 4 pattern.)

Mark the pattern of eights on the same chart, using a different colour or shape.

What do you notice about the pattern now? (Some numbers are in both the 4 and 8 patterns.)

List the numbers that are multiples of both 4 and 8: 8, 16, 24, 32, 40 . . .

Is every multiple of 8 also a multiple of 4? Is every multiple of 4 also a multiple of 8? (Every number in the 8 pattern is in the 4 pattern. But not every number in the 4 pattern is in the 8 pattern.)

Variation

Repeat the activity for multiples of 3 and 6, and 5 and 10.

Threes, sixes and nines

Materials

- a blank hundreds chart

Show the chart and mark this pattern:

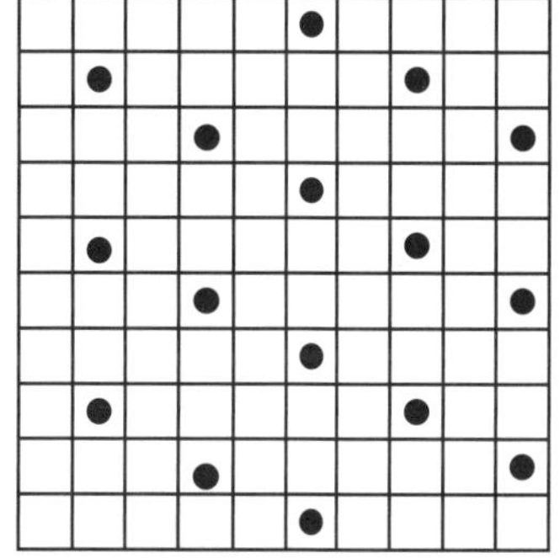

What can you tell me about the pattern? (It is going up by 6 each time. It is a counting by 6 pattern.)

Mark the pattern of threes and nines on the same chart, using different colours or shapes.

What do you notice about the pattern now? (Some numbers are in all three patterns.)

List the numbers that are multiples of 3, 6 and 9: 18, 36, 54, 72, 90

What can you say about these numbers? (They are going up by 18.) Can all multiples of 9 be divided by 3? (Yes) Can all multiples of 9 be divided by 6? (No)

Calculator patterns

Materials

- a class set of calculators
- pen and paper

Ask a student to suggest a 3-digit number, for example, 328.

Can you count forwards by 20s from 328?

Tell the students that you will check their count using the calculator. Enter [3] [2] [8] [+] [2] [0] [=] and then continue to press [=] so that the calculator generates a pattern that counts by 20. Ask students to input any 3-digit number on their calculator and practise counting forwards by 20s. Limit the activity to 10 presses.

Variation

Repeat the activity counting by 200 and 2000.

Higher terms

Show this number pattern:

6 12 18 24 30

What is the next term in the pattern? How did you work that out? (It is the pattern of sixes. You add 6 each time.) What is the tenth term in the pattern? (For the second term, you multiply 2 sixes. For the tenth term, it must be 10 sixes.) What is the hundredth term in the pattern?

Discuss the students' mathematical thinking strategies.

Missing patterns

Show these frames with the numbers:

Can you see a pattern in each frame? (There is a difference of 40. The number in between must be half that . . .20.)

	1145		1185		1225	
870		970		1070		
	2100		2150	2175		

Complete each frame as the students supply the answers. Ask students to count the numbers in each pattern, forwards and backwards.

Tricky patterns

Show these frames with the numbers:

Can you see a pattern in each frame? (There is a difference of 200. The number in between must be half that . . . 100.)

	4531		4731		4931	
764		1164		1564		
	3700		4700		5700	

Complete each frame as the students supply the answers. Ask the class to count the numbers in each pattern, forwards and backwards.

Tables patterns

Materials

- a multiplication grid

Show the grid and discuss any patterns observed by the students:

(Start at 40 on the bottom row and go up diagonally to the right. It goes up by 5, then 3, then 1, 1 again, 3 then 5. You can count by sixes along a row or down a column. I can see pairs of numbers on each side. 4 threes are 12 and 3 fours are 12.)

×	1	2	3	4	5	6	7	8	9	10
1	1	2	3	4	5	6	7	8	9	10
2	2	4	6	8	10	12	14	16	18	20
3	3	6	9	12	15	18	21	24	27	30
4	4	8	12	16	20	24	28	32	36	40
5	5	10	15	20	25	30	35	40	45	50
6	6	12	18	24	30	36	42	48	54	60
7	7	14	21	28	35	42	49	56	63	70
8	8	16	24	32	40	48	56	64	72	80
9	9	18	27	36	45	54	63	72	81	90
10	10	20	30	40	50	60	70	80	90	10

Discuss how the opposite pairs of table facts illustrate the commutative property, for example, $6 \times 4 = 4 \times 6$. Check whether the students can identify the square number pattern running diagonally across the table.

Tables chart

Materials

- a multiplication grid

Show a multiplication grid and discuss any patterns observed by the students.

How could you find the answer to 2 × 4 using this tables chart?

Explain that when multiplying 2×4, the product 8 is found at the intersection of the second row and fourth column. The commutative law for multiplication can be investigated by locating 4×2 in a similar fashion.

How could you find the answer to 10 ÷ 5?

To find the missing factor for $10 \div 5$, look for the product 10 in the fifth row, then go up the column to find the missing factor 2 at the top. Continue relating multiplication and division facts until the students can remember this pattern.

Circular patterns

Materials

- 3-pattern wheel

Show a pattern wheel.

Mark the pattern of twos around the circle to form a pentagon:

Count the numbers in the pattern: 0, 2, 4, 6, 8, 0.

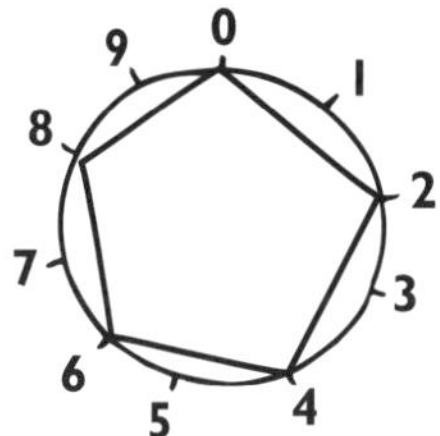

Mark the pattern of fours, continuing around the circle to form a 5-pointed star:

Count the numbers in the pattern: 0, 4, 8, 12, 16, 0.

What do you notice about the two patterns? (The patterns of 2 and 4 are the same. It is because each pattern has a 2, 4, 6 and 8 somewhere in the ones place of the numbers in the pattern.)

Repeat the activity for the pattern of sixes.

Complex patterns

Show this pattern wheel:

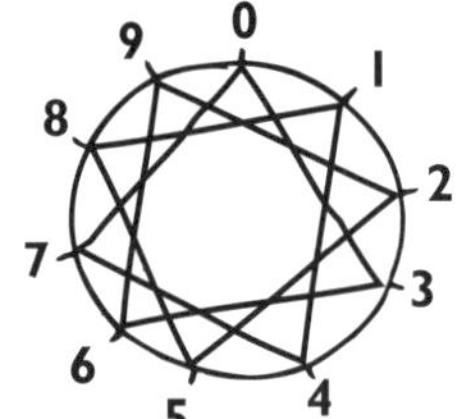

Count the numbers in the pattern:
0, 3, 6, 9, 12, 15, 18, 21, 24, 27, 0.

Mark the pattern of sevens around the circle:

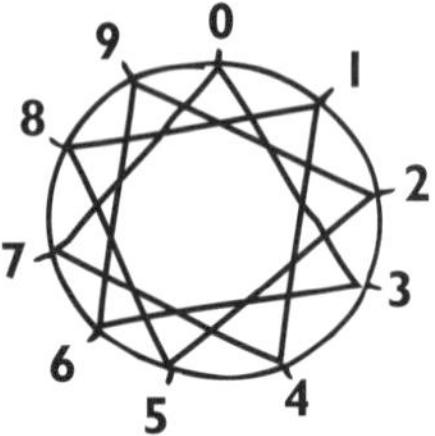

Count the numbers in the pattern:
0, 7, 14, 21, 28, 35, 42, 49, 56, 63, 0.

What do you notice about the patterns of threes and sevens? (They are the same because each pattern has all the digits from 0 to 9 somewhere in the ones place of its numbers.)

Balancing multiplication

Show students a pan balance and ask them to imagine using it with multiplication facts. On one side of the balance show the multiplication number fact: 8 × 3

Which other multiplication facts will make the balance level? (You must get 24 . . . 2 twelves are 24. So, 12 multiplied by 2 is 24 . . . 2 × 12.)

Show the equivalent number facts on the opposite side of the balance. Emphasise the idea of balance between both sides of an equation. The equals sign indicates 'is the same as', as opposed to a signal to perform an operation (+, –, ×, ÷).

Repeat the activity for: 6 × 6, 10 × 6.

Balancing division

Show students a pan balance and ask them to imagine using it with division facts. On one side of the balance show the division number fact: **63 ÷ 7**

Which other division facts will make the balance level? (You must get a 9. So, 9 fours are 36 . . . 36 divided by 4 is 9.)

Show the equivalent number facts on the opposite side of the balance. Emphasise the idea of balance between both sides of an equation. The equals sign indicates 'is the same as', as opposed to a signal to perform an operation (+, –, ×, ÷).

Repeat the activity for: **56 ÷ 7** **45 ÷ 9**

Variation

Challenge the students to write their own facts for both sides of the balance, with mixed operations, for example, 72 ÷ 9 = 2 × 4; 6 × 6 = 72 – 36.

Garden number sentences

Ask the students to think of a number sentence and answer to match this word problem:

A gardener had 8 rows of 12 orange trees, how many orange trees did they have? (8 × 12 = 96) How many rows could the gardener make if they had 72 orange trees? (That is 72 divided by 12.)

If the gardener wanted to plant 64 lemon trees, how might they arrange these? (They might do 8 rows of 8. Or they could do 16 rows of 4. They could do 2 rows of 32.)

Students can work in pairs to create their own word problems for a variety of number sentences.

Missing addition (1)

Show these number sentences:

______ = 9 + 18 **5 + ______ =14** **______ + 6 = 24**

How would you solve these number sentences? (That means 5 plus something equals 14. You can subtract 5 from 14 . . .)

Discuss the students' strategies. Repeat the activity using similar examples. Note that students will often give answers of 27, 19 and 30 to these examples. This is an indication that they see '=' as an indicator to simply perform an operation, rather than expressing an equivalent number relationship. Use examples that emphasise the inverse operations and repeat the idea of the equation often.

Missing addition (2)

Show these number sentences:

______ = 18 + 36 **23 + ______ = 49** **______ + 64 = 128**

How would you solve these number sentences? (23 add something equals 49. 20 makes 43 and 6 more makes 49 . . . 26.)

Discuss the students' strategies. If necessary, relate explanations to a number line segment. Note that students often see '=' as an indicator to simply perform an operation. Solving number sentences in the reverse form and with missing addends helps students to understand the idea of an equation.

Missing subtraction (1)

Show these number sentences:

______ = 17 – 6 **15 – ______ = 7** **______ – 6 = 21**

How would you solve these number sentences? (Something take away 6 is 21. It must be 6 more than 21 – 27. 15 minus something = 7. 8 and 7 is 15 so the answer is 8.)

Discuss the students' strategies. Repeat the activity using similar examples.

Missing subtraction (2)

Show these number sentences:

______ = 78 – 39 **86 – ______ = 21** **______ – 105 = 225**

What do each of these number sentences say? How would you solve them? (86 subtract something equals 21. I can count back from 86 to 21. If you take 105 from a number, you get 225.)

Discuss the students' strategies. Repeat the activity using similar examples.

UPPER – YEARS 5 AND 6

Number and Algebra

NUMBER AND PLACE VALUE

Year 5

Five-digit breakdown

Materials

- Base 10 materials

Show this number: **24 723**

Can you imagine what this number looks like using Base 10 materials? (You have got 247 hundreds.) How many ten thousands? Thousands? (There are 24 thousands. There would be 24 one thousand blocks.) Hundreds? Tens? (I can imagine 2 long 10 000 blocks. That is 20 000.) How many ones in 24 723? (You would need 24 723 shorts. That is a lot.)

Discuss the students' mental imagery of these numbers. Use the Base 10 materials to demonstrate the different representations of 24 723.

Place it

Show this number line segment from 10 000 to 11 000, with intervals of **200**:

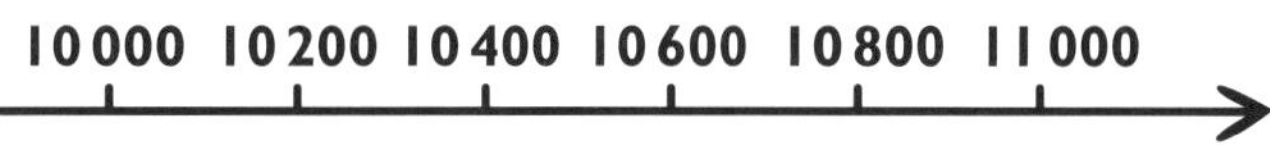

Show these numbers: **10 300, 10 250, 10 990, 10 660, 10 410**

Who can show me the position of these numbers on the number line? Why are they placed there?

Invite different students to point out the position of the numbers on the number line. Discuss the students' responses.

Number line intervals

Show an open number line segment from 50 000 to 60 000:

50 000 60 000

Can you get from 50 000 to 60 000 in two equal jumps? What would be the number in between? (It is the halfway mark – 55 000.)

Discuss and point to the midpoint on the number line. Repeat the activity for four jumps. Count forwards and backwards in intervals of 2500 along the line.

Guess, check and refine

Show an open number line segment from 32 500 to 33 500, with intervals of 100:

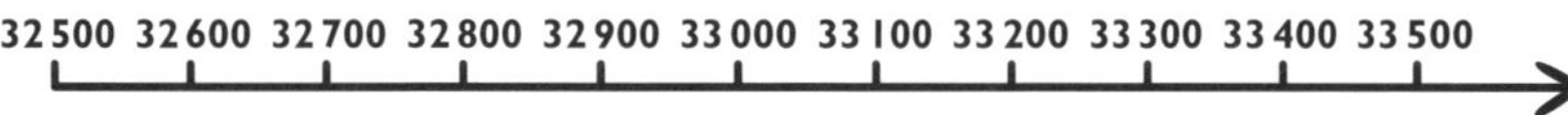

I am thinking of a number between 32 500 and 33 500. Can you guess my number? (Is the number larger than 32 700? Is it between 33 300 and 33 500?)

Ask the students to guess the number in turn. Respond with yes/no statements such as, 'No, the number is not between 33 300 and 33 500'. Mark each suggestion on the number line segment until the chosen number is guessed.

Variation

Play the game within different ranges. Allow the students to pose their own numbers for increased motivation.

Five-digit wipe out

Materials

- a class set of calculators

Enter [7] [4] [8] [9] [3] on the calculator.

How could you change the 4 in the display to a zero? Why did you subtract 4000? (It is because the 4 is in the thousands column. It stands for 4000.)

Discuss the students' responses to these questions:

How could you change the 8 in 74 893 to a zero? How could you remove the 7? How could you get the answer 893? How could you get a remainder of 93?

Variation

Repeat the activity with other 5- and 6-digit numbers.

Five-digit shuffle

Materials

- a set of number cards from 0 to 9

Show number cards for:

How could you arrange the numeral cards to make the largest number possible? Why did you arrange the cards that way? (Ten thousands are bigger, so you put the 7 in that place.) What would be the second-largest number you could make? Why is the second-largest number not 57 430? (If you change the ten thousands, the number becomes much smaller. If you change the ones and tens you get 75 403. That is much closer to 75 430.)

Repeat the activity, with students choosing their own number cards.

Five-digit draw

Materials

- two sets of number cards from 0 to 9
- two 6-column place value charts

Divide the class into two teams. Explain to the students that you are playing a game where the aim is to draw digit cards and make the lowest 5-digit number possible.

Rules for 5-digit draw

1. Choose students from each team in turn, to select a number card.
2. Students place their cards in the column of their choice on their team's place value chart.
3. After five draws, the team with the highest number is the winner.

Variation

Change the aim of the game to make the highest 5-digit number. Repeat the activity for 6-digit numbers.

Higher or lower

Materials

- a collection of 5-digit numbers on cards

Select a student to choose one of the cards and to keep the nuber secret. Ask the students to guess the number. The student who chose the card can only respond with 'higher' or 'lower' until the chosen number is guessed.

Variation

Record the students' responses progressively on a number line as the game is played. Repeat the game for 6-digit numbers.

Reading numbers

Show these numbers:

114 398 **364 392** **286 205** **818 960**

Can you write these numerals in words? (That is two hundred and eighty-six thousand, two hundred and five and so on.)

Skipping numerals

Show these numbers:

15 632 **58 900** **12 007** **109 019**

Can you tell me how to read these numbers?

Discuss the students' responses. Remind students of the convention in naming numbers containing zeros: the place value of the column containing zero is not said, for example, 12 007 is read as 'twelve thousand and seven', not as 'twelve thousand zero hundreds, zero tens and seven'.

Arrays and factors (1)

Materials

- 72 counters

Arrange the counters at random.

How many different arrays could we make using 72 counters? (12 groups of 6 make 72. It is 12 by 6. I have got 4 eighteens.) How many counters long and wide will each array be? (1 × 72, 2 × 36, 3 × 24, 4 × 18, 6 × 12)

Encourage students to work in pairs to discuss arrays of different sizes.

Can you tell me the factors of 72? (1, 2, 3, 4, 6, 12, 18, 24, 36, 72)

Arrays and factors (2)

Materials

- 96 counters

Arrange the counters at random.

How many different arrays could we make using 96 counters? (I have got 4 × 24 and 12 × 8, 16 groups of 6 make 96.) How many counters long and wide is each array? (1 × 96, 2 × 48, 3 × 32, 4 × 24, 6 × 16, 12 × 8)

Encourage students to work in pairs to discuss arrays of different sizes.

Can you tell me the factors of 96? (1, 2, 3, 4, 6, 8, 12, 16, 24, 32, 48, 96)

Variation

Students can investigate the factors of other numbers in the range from 0 to 200.

Factor trees

Show this factor tree:

What number should be at the top of the factor tree? (48) How did you work it out? (I used 12 × 4.) Can you add other multiples to complete this factor tree? (3 × 4 = 12, 2 × 2 = 4)

12 4

Repeat the activity for these numbers: 18, 30, 64, 96.

Divisibility rules

Show these numbers: **275, 628, 470, 312, 724, 980**

Which of these numbers are divisible by 2, 5 and/or 10? (470 and 980 are divisible by 10 because they end in 0; 470 and 980 can also be divided by 5; all numbers ending in 2, 4, 6, 8 and 0 are even numbers so can be divisible by 2.)

Students can work in pairs and report back to the class. Make a list of some suggested divisibility rules for 2, 5 and 10.

If the sum of the digits in a number is divisible by 3, then that whole number will be divisible by 3, for example, 531 = 5 + 3 + 1 = 9, 9 ÷ 3 = 3, so 531 is divisible by 3; if the last two digits of a number are divisible by 4, the whole number will be too, for example, 312 = 12 ÷ 4 = 3, so 312 can be divisible by 4.

Variation

Ask the students to suggest some 2-, 3- and 4-digit numbers that are divisible by 3 and 4.

Everyday estimations (1)

Show these measurements: **19.2°** **18.5 km** **791 m²** **43 000 L**

What would appropriate estimates be for a maximum temperature of 19.2°? (That is about a 20° maximum.) A bushwalk of 18.5 km? (It is a walk of about 20 km.) A block of land of 791 m²? (The block is about 800 m².) A swimming pool that holds 45 000 L? (The pool holds over 40 000 L.)

Discuss the students' responses and the level of precision required for estimates in daily life.

Larger estimations

Show these numbers: **1050** **54 236 489 900** **19 990**

What would appropriate estimates be for a bike that costs $1050? (I need to save about $1000 for that bike.) A one-day cricket crowd of 54 236? (That is a crowd of about 55 000.) A house that sold for $489 900? (The house cost half a million dollars.) A car that costs $19 990? (The car costs $20 000.)

Discuss the students' responses and the level of precision required for estimates in daily life.

Everyday estimations (2)

Show these amounts:

3.89 million people **24.263 tonnes** **17 915 m²** **# 483 241**

What would appropriate estimates be for a population of 3.89 million people? (That is about 4 million people.) A truck that weighed 24.263 tonnes? A shopping centre covering an area of 17 915 m²? A lotto prize of $483 241? (The person won nearly half a million dollars.)

Discuss the students' responses and the level of precision required for estimates in daily life. Ask the students to suggest some situations of their own where the use of estimates is appropriate.

Three-addend estimates

Show this number sentence: **393 + 205 + 211 =** ______

How would you estimate the answer to this number sentence? (That is about 400 add 200 add 200. About 800 altogether.)

Discuss the students' estimation strategies. After estimates have been established, encourage the students to calculate the answer. Mental, written or calculator calculation may be used.

Do you think your answer is reasonable or not? Why? (I calculated 809. It is very close to the estimate so I think I am right.)

Repeat the activity for these number sentences: **176 + 288 + 112** **990 + 437 + 221**

Four-addend estimates

Show this number sentence: **435 + 637 +172 + 315 =** ______

How would you estimate the answer to this number sentence? (400 plus 600 equals 1000. Add 200 plus 300. About 1500 altogether.)

Discuss the students' estimation strategies. After estimates have been established, encourage the students to calculate the answer. It is appropriate to use a calculator for calculations with more than three addends.

Do you think your answer is reasonable or not? Why? (My calculator answer is 1559. It is reasonably close to the estimate.)

Repeat the activity for: **176 + 288 + 112** **990 + 437 + 221**

Front-end estimation

Show this number sentence: **4517 + 3216 + 6376 =** ______

How would you estimate the answer to this number sentence? (You just add up the thousands digits. It is 5000 plus 3000 plus 6000 . . . 14 000.)

Discuss the students' estimation strategies. Encourage the students to then calculate the answer. It is appropriate to use a calculator for these types of examples.

Do you think your answer is reasonable or not? Why? (My estimate was 14 000. The calculator says 14 109. That looks correct.)

Repeat the activity for: **4348 + 2782 + 5384 =**______

Rounding off

Show these number sentences:

5473 – 2287 = ______ **3456 + 2318 =** ______

How would you estimate the answer to these number sentences? (You can round off the hundreds. It is 54 hundred minus 22 hundred, that is 32 hundred.)

Discuss the students' estimation strategies. Encourage students to then calculate the answer.

Repeat the activity for:

3505 – 1482 **4022 + 3855**

Estimation by half

If you went to the shop and you wanted to buy 5 items for $5.48 each, about how much money would you need to have? (I think it is $6 × 5 that is $30. That is about $5 × 5 equals $25. If the lower estimate is $25 and the upper estimate is $30, I would say the best estimate is $27.50.)

Discuss the students' estimation strategies. Encourage students to then calculate the answer. Recall situations in daily life where estimates rather than exact calculations are used.

Year 6

Seven-digit wipe out

Materials

- a class set of calculators

Enter these numbers on the calculator: 1 3 7 5 2 8 4

How could you change the 3 in the display to a zero? Why did you subtract 300 000? (It is because the 3 is in the hundred thousands column. It stands for 300 000.)

Discuss the students' responses to these questions:

How could you change the 5 in 1 375 284 to a zero? How could you remove the 1? How could you get the answer 284? How could you get an answer of 1 000 000?

Variation

Repeat the activity with numbers of any size. Encourage the students to develop questions of their own.

Seven-digit shuffle

Materials

- a set of number cards from 0 to 9

Show the number cards for: 9 6 3 4 5 0 7

How could you arrange the number cards to make the largest number possible? Why did you arrange the cards that way? What would be the second-largest number you could make? Why is the second-largest number not 7 965 430? (You need to change the ones and the tens digits to get the second-largest number.)

Discuss the students' responses.

Variation

Ask seven students to randomly select a number card each. Repeat the activity.

Seven-digit draw

Materials

- two sets of number cards from 0 to 9
- two 7-column place value charts

Divide the class into two teams. Explain to the students that you are playing a game where the aim is to draw digit cards and make the lowest 7-digit number possible.

Rules for 7-digit draw

1. Select a student from each team in turn, to choose one number card each time.
2. Students place their cards in the column of their choice on their team's place value chart.
3. After four draws, the team with the lowest number is the winner.

Variation

Repeat the activity so that the aim is to make the highest 7-digit number. Play the game with numbers of any size.

Spot the largest

Show these series of numbers:

10 111	**11 001**	**11 010**	**11 011**
618 542	**615 802**	**618 852**	**562 546**
1 222 387	**1 113 907**	**1 208 000**	**1 213 097**
2 017 905	**2 071 606**	**2 070 696**	**2 007 259**

Which is the largest number in each group? How do you know?

Discuss the relative sizes of the numbers in terms of their place value. Circle the largest number in each series. Repeat the activity for the smallest number in each group.

Expanded form

Materials

- a set of number cards from 0 to 9

Choose seven students to select one number card each. Arrange the number cards at random to form a 7-digit number.

How could we record this number using its place value?

Show the expanded form of the number, for example:

2 186 337 = 2 000 000 + 100 000 + 80 000 + 6000 + 300 + 30 +7

Repeat the activity, showing other 7-digit numbers in the expanded form.

Order for five

Materials

- 5 cards

Show these numbers on the cards:

Choose five students to select one card each.

3 260 090	3 259 996	2 360 169	2 660 196	3 259 096

Can you order these numbers from lowest to highest? (3 260 090 is highest as it has the highest number in the ten thousands place.) What is the second-highest number in the sequence? How do you know? Which number is in the middle?

Discuss the relative sizes of the numbers in terms of their place value. Repeat the activity for other sets of numbers of any size.

Write and read

Show this number: **274 428 987**

How do you read this number? (274 428 987. That is 274 million, 428 thousand and 987.)

Develop the convention of leaving spaces (not commas) between each group of three digits when writing numbers larger than four digits.

Explain to the students the repeated pattern of ones, tens and hundreds used when naming numbers according to place value:

hundreds	tens	ones	hundreds	tens	ones	hundreds	tens	ones
millions			thousands			ones		
100 millions	10 million	millions	100 thousands	10 thousands	thousands	hundreds	tens	ones
2	7	4	4	2	8	9	8	7

More than millions

Show this number: **1 000 000 000**

How do you read this number? How do you know this is 1 billion? (That is 1 with 9 zeros after it. It is 1 billion. 1 million has 6 zeros after the first digit.)

Encourage students to share their responses and show the number using spaces. Discuss the naming of the number according to the place value of its digits. Explain how the convention of leaving spaces (not commas) between groups of 3 digits makes the naming easier.

Variation

Encourage students to research multi-digit numbers. Discuss how to read and write the numbers with the students.

Place it

Show an open number line segment from 200 000 to 500 000, with no intervals:

200 000 **500 000**

Show these numbers:

222 000 **420 000** **290 000** **315 000** **351 000**

Who can show me the position of these numbers on the number line? Why are they placed there?

Invite different students to mark the position of the numbers on the number line. Discuss the students' suggestions.

Number line intervals

Show an open number line segment from 0 to 500 000, with no intervals:

0 **500 000**

Can you get from 0 to 500 000 in 10 equal jumps? (We could try 50 000. After 5 jumps we are at the halfway mark of 250 000.) What would be the numbers in between?

Discuss the students' suggestions and mark the intervals on the number line. Repeat the activity for 20 jumps. Count orally forwards and backwards in intervals of 25 000 along the line.

Ten thousands jumps

Show an open number line segment from 600 000 to 800 000, with intervals of 10 000:

Invite the students to count by 10 000s from 600 000 to 800 000, forwards and backwards.

If I started at 650 000, how many jumps of 10 000 would it take to get to 800 000? (150 000 is the difference. You need 15 jumps of 10 000.) If I was at 740 000 and made 3 more jumps of 10 000, where would I be? How many jumps of 10 000 from 600 000 to 800 000?

Discuss the students' suggestions and mark the jumps on the number line.

Five jumps

Show an open number line segment from 0 to 1 000 000, with no intervals:

0 **1 000 000**

Can you get from 0 to 1 000 000 in 5 jumps? What would be the numbers in between? (You can use equal jumps or unequal jumps. I will go 50 000, 300 000, 500 000, 700 000 and 1 000 000.)

Discuss the students' suggestions and mark the jumps on the number line. Count forwards and backwards orally using the intervals suggested by the students.

Variation

Repeat the activity using various numbers of unequal or equal jumps.

Making a million

Materials

- about 30 Base 10 large cubes
- a 1 m^3 kit or 12 one-metre rulers and duct tape

Can you imagine what 1 million would look like using Base 10 materials?

Allow time for students to discuss their mental imagery of 1 million.

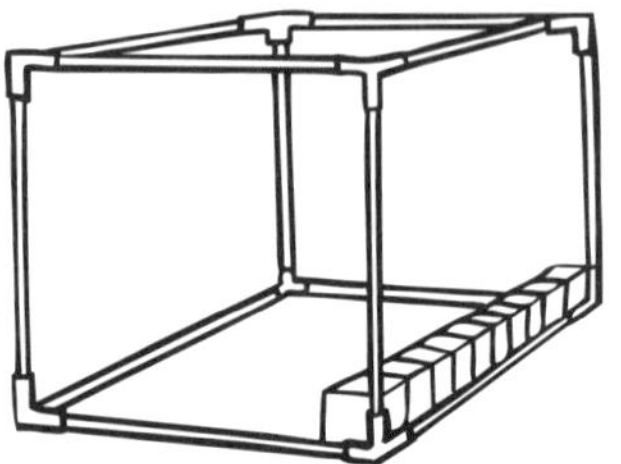

How long is the structure? How wide? How tall?

Construct the model of 1 million with the students. A skeleton model of 1 m^3 can be used. Place rows of Base 10 large cubes along the length, width and height of the metre frame.

Students may investigate other representations of 1 million, for example, they could use the squares on sheets of 2 mm grid paper.

A million questions

Materials

- the model of 1 million from the previous activity

Examine the model of 1 million with the students.

How many thousands in a million? (There are 10 by 10 by 10 thousands in 1 million. That is 1000 thousands.) How many 10 thousands in a million? How many 100 thousands in a million?

Discuss the students' responses and encourage them to demonstrate their answers using the Base 10 material. Develop the understanding that the model contains 1 million 'shorts' or 'minis'.

Links with volume may be made using this activity. The model constructed occupies 1 m^3 and contains 1 million cm^3.

Primes and composites (1)

How many different arrays can you make for each of the numbers from 30 to 40? How many prime numbers are there between 30 and 40? Which number between 30 and 40 has the most factors?

Show the numbers from 30 to 40 in a table. Consider each of the numbers in turn with the students and record the students' responses.

Number	30	31	32	33	34	35	36	37	38	39	40
No. of arrays	4	1	3	2	2	2	5	1	2	1	4
No. of factors	8	2	6	4	4	4	10	2	4	4	8
Prime or composite	C	P	C	C	C	C	C	P	C	C	C

Remind students that prime numbers have only one possible array and two factors (the number itself and 1) and composite numbers have more than one possible array and more than two factors. (Note that 1 is not a prime number because it has only one factor, itself.) Repeat the activity for 40 to 50 and 50 to 60.

Variation

Investigate prime and composite numbers from 60 to 100.

Primes and composites (2)

How many different arrays can you make for each of the numbers from 2 to 12?

Show the numbers from 2 to 12 in a table. Consider each of the numbers in turn with the students and record their responses.

Number	2	3	4	5	6	7	8	9	10	11	12
No. of arrays	1	1	2	1	2	1	2	3	2	1	3
No. of factors	2	2	3	2	4	2	4	3	4	2	6
Prime or composite	P	P	C	P	C	P	C	C	C	P	C

Remind students that prime numbers have only one possible array and two factors (the number itself and 1) and composite numbers have more than one possible array and more than two factors. (Note that 1 is not a prime number because it has only one factor, itself.)

Variation

Investigate prime and composite numbers from 13 to 30.

Eratosthenes' sieve

Materials

- a hundreds chart
- coloured pencils

Show a hundreds chart:

1	2	3	4	5	6	7	8	9	10
11	12	13	14	15	16	17	18	19	20
21	22	23	24	25	26	27	28	29	30
31	32	33	34	35	36	37	38	39	40
41	42	43	44	45	46	47	48	49	50
51	52	53	54	55	56	57	58	59	60
61	62	63	64	65	66	67	68	69	70
71	72	73	74	75	76	77	78	79	80
81	82	83	84	85	86	87	88	89	90
91	92	93	94	95	96	97	98	99	100

Ask pairs of students to cross out multiples of 2, 3, 5 and 7. Use a different colour for each set of multiples.

What do you notice about the chart? Which type of numbers has been crossed out? Which type of numbers remains? Why were those numbers not crossed out? How many common multiples can you find?

Discuss the students' mathematical thinking in relation to these questions. As a class, research and discuss the work of Eratosthenes with prime numbers and mathematical geography.

Product and sum

I am thinking of two numbers whose sum is 32 and product is 128. What are my numbers? (16 and 16 is 32, but 16 squared will be too big. I will try 28 and 4, 28 + 4 = 32 and 28 × 4 = 128, it is 28 and 4.)

No. 1	No. 2	+	×
16	16	32	256
20	12	32	220
28	4	32	128

Discuss the students' strategies, especially the guess–check–refine strategy. A table may also be useful to help organise students' thinking, and calculators may be used.

Repeat the activity.

My difference is 8 and my quotient is 3. What are my numbers? (12 and 4)

Ensure that the students are familiar with the mathematical terms 'sum', 'difference', 'product' and 'quotient' in relation to the four operations.

Triangular patterns

Materials

- counters

Show this pattern:

What is the next shape in the pattern? How did you work that out? (It is a pattern of triangles. You add a new row with an extra one each time.)

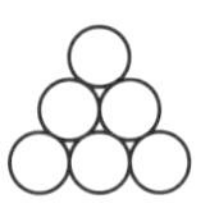

Show the first five numbers in the pattern as 1, 3, 6, 10 and 15. Focus on the differences between those numbers: 2, 3, 4, 5.

What are the next 5 numbers in the pattern? (21, 28, 36, 45, 55) What are the differences between the numbers?

Square numbers

Show this pattern:

What is the next shape in the pattern? How did you work that out? (It is a pattern of squares. You multiply the number by itself.) What do we call these numbers? (Square numbers)

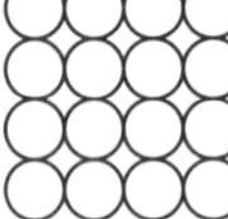

Explain to students that a number multiplied by itself is a square number and can be depicted as a square array.

What is the next number in the pattern?

Addition and subtraction

Year 5

Magic squares

Show this 4 × 4 grid:

18	4	5	15
7	13	12	10
11	9	8	14
6	16	17	3

What do you notice about the numbers in this grid? (It totals 42, up, down or diagonally. If you add them, they always add up to the same number.) How many combinations of 42 can you find? (10)

Discuss the students' solutions and encourage the students to construct their own magic squares (see the next activity).

Making magic

Would you like to learn how to construct your own magic square?

Show this 4 × 4 grid:

1	2	3	4
5	6	7	8
9	10	11	12
13	14	15	16

Show the students these steps to make a magic square.

1. Transfer the positions of the numbers in the opposite corners, that is, 1 and 16, 4 and 13.

2. Repeat the process for the four inner numerals, that is, 6 and 11, 7 and 10.
3. Copy the remaining numbers without changing their positions.

What is the sum of the rows, columns and diagonals in this square? (34)

16	2	3	13
5	11	10	8
9	7	6	12
4	14	15	1

Students can begin their first square with a single-digit number of their choice in the upper left corner, then fill the remaining squares with consecutive numbers, and follow the steps to create their magic square.

Variation

Students can exchange their magic squares and check their constructions by calculating answers. They can also research the history of the invention of magic squares.

Three-digit addition

Show this number sentence: **627 + 111 =** ______

How would you solve this problem? (627 add 100 equals 727, add 11 equals 738. There is an easy way, all the digits go up by 1. You add the hundreds first, that is 700 plus 30 plus 8.)

Discuss the range of strategies that students provide and record them, for example:

Jump 627 + 100 + 10 + 1

Split 600 + 100 + 20 + 10 + 7 + 1

Repeat the activity for: **374 + 205 =** ______ **435 + 309 =** ______

Three-digit trading

Show this number sentence: **149 + 258 =** ______

What is the answer to this number sentence? Can you tell me how you worked it out? (Add 1 onto 149 and take one off 258, that is 150 add 257 . . . 357 . . . 407. You add the hundreds, tens and then the ones, so that is 300, 9 tens and 17 . . . 407. You can add 1 to 149 to make it 150, you can add 2 to 258 to make it 260, then 150 plus 260 is 410, take away 3 gives 407.)

Discuss and list the different strategies used by students. Encourage as wide a range of strategies as possible.

Compensation addition

Show this number sentence: **439 + 198 =** ______

What is the answer to this number sentence? Can you tell me how you worked it out? (You add 2 to 198 to make it 200; 439 add 200 is 639, take away 2 gives 637.)

Discuss the students' responses. Focus on the compensation strategy of rounding a number to obtain the solution.

Repeat the activity for these number sentences:

384 + 497 = ______ **548 + 295 =** ______

Note that when you are discussing a particular strategy with the students, it is never the intention to prescribe a fixed approach. The aim is to develop number sense by modelling a range of mental strategies from which students can choose.

Compensation subtraction

Show this number sentence: **645 – 139 =** ______

What is the answer to this number sentence? Can you tell me how you worked it out? (If you take away 140 you get 505, add 1 and it is 506.)

Discuss the students' responses. Focus on the compensation strategy of rounding a number to obtain the solution.

Repeat the activity for these number sentences: **792 – 408 =** ______ **876 – 267 =** ______

384 + 497 = ______ **548 + 295 =** ______

Note that when you are discussing a particular strategy with the students, it is never the intention to prescribe a fixed approach. The aim is to develop number sense by modelling a range of mental strategies from which students can choose.

Changing order

Show these number sentences:

18 + 176 + 282 + 4 = ______ **350 + 401 + 50 + 19 =** ______

21 + 19 + 6 + 894 = ______ **187 + 9 + 13 + 691 =** ______

Can you see a quick way to add these?

Encourage the students to search for compatible numbers, adding them in any order.

Variation

Ask the students to write some similar addition sentences of their own. Extend to five addends when the students are ready.

Fill the blanks

Show this grid:

How can you work out the missing numbers? Did you add or subtract to fill in the blanks in each row?

Discuss the strategies suggested by the students.

T Th	Th	H	T	O	Total
10 000	4000	200	70	6	
	1000	400	40	3	21 443
30 000			90	1	32 691
	5000	300			55 372
70 000					70 084

Number line patterns

Show this number: **4950**

Can you tell me two numbers that add up to this number? Can you suggest other combinations? (2400 and 2550 equals 4950. I think you should plus 1000 and 3950. You can add 4000, 900 and 50.)

Show a number line from 0 to 5000, with intervals of 100:

Mark the number line to illustrate students' suggestions. Repeat the activity for similar 4-digit numbers.

Find the difference

Show these numbers: **2150, 3200**

How could you find out the difference between 2150 and 3200? (You can jump 50 up to 2200, 500 up to 2700 and then another 500 to 3200.)

Discuss the students' responses. If necessary, introduce the idea of jumps on the number line to obtain the answer. Ask the students to demonstrate different jumps that could be made. Repeat the activity for the difference between:

3325 and 4300 **4375 and 5125** **6190 and 7705**

Find the number

Show this number: **620**

This number is the difference between two numbers. Can you tell me what the two numbers are? (If you take 1000 from 1620, you get 620. If you subtract 620 from 1240 the answer is 620.)

Discuss the students' responses. Repeat the activity for other 3- and 4-digit numbers.

Three for 20 000

Show this number sentence: ______ **+** ______ **+** ______ **= 20 000**

Can you tell me 3 numbers that add up to this number? (9400 plus 600 plus 10 000 equals 20 000. 9000 add 5500 and 5500 makes 20 000.)

0 **20 000**

Discuss the students' responses and demonstrate their strategies using jumps on a number line segment from 0 to 20 000.

Two for 100 000

Show this number sentence: ______ **+** ______ **= 100 000**

Can you tell me 2 numbers that add up to this number? (45 000 plus 55 000 equals 100 000.

9000 add 91 000 is 100 000.)

0 **100 000**

Discuss the students' responses and demonstrate their strategies using jumps on a number line segment from 0 to 100 000.

Subtract two for 10 000

Show this number sentence: ______ **–** ______ **= 10 000**

Can you tell me two numbers that have a difference of 10 000? (19 428 subtract 9428 equals 10 000. 21 000 take away 11 000 equals 10 000.)

0 **100 000**

Discuss the students' responses and demonstrate their strategies using jumps on a number line segment from 0 to 100 000.

Comparing numbers

Show this list of numbers: 700, 2892, 39, 441, 14 400, 7200

How does the number 1439 compare to the rest of these numbers? (2892 is roughly double 1439. 700 is about half of 1439. If you take away 1400 you get 39. 14 400 is about 10 times 1439.)

Discuss the students' responses. Practise similar number comparisons often to develop students' estimation skills.

Roman dates

Sheri and Karly watched a movie. When the credits rolled at the end of the movie they saw the Roman numerals MCMLXXXVII. What did this mean? (I think M stands for 1000. C before M must mean 900. That is 1000 + 900 + 50 + 30 + 5 + 2. It makes 1987 altogether.)

Discuss the students' responses.

Is it easier to write this number in Roman numerals or modern numerals? Why?

Discuss the advantages of our modern Hindu–Arabic system over the Roman system, that is, it has place value, zero as a place holder and makes written calculation in columns possible. Students can research the history and traditional uses of Roman numerals.

Year 6

Magic calendar

Show this 3 × 3 grid:

5	6	7
12	13	14
19	20	21

Tell the students that the grid is like a section from a calendar.

If you add the columns, rows or diagonals in the array, what do you notice? How could you explain these answers? (The difference between the totals of the rows is 21. If you add the columns, the difference between each column total is always 3. If you add the middle column and the middle row, the sums are equal.)

Discuss the students' responses.

Variation

Repeat the activity with a different section of a calendar. The calendar is a counting chart with a row increase of 1 and a column increase of 7. The total of 3 consecutive columns will have a difference of 3. The totals of 3 consecutive rows will have a difference of 21. Adding the diagonals will produce the same answer in both directions.

More magic

Show this 4 × 4 grid:

24	10	11	21
13	19	18	16
17	15	14	20
12	22	23	9

What do you notice about the numbers in this grid? (If you add them up, they always add up to the same number. The total is 66 – up, down or diagonally.) How many combinations of 66 can you find?

Discuss the students' responses and encourage them to make their own magic squares.

Missing values

Show this table:

H Th	T Th	Th	H	T	U	Total
10 000	20 000	7000	300	80	4	
		2000	400	40	6	212 446
	40 000			10	1	443 711
		6000	300			376 376
	70 000					970 010

How can you work out the missing numbers? Did you add or subtract to fill in the blanks in each row?

Discuss the strategies suggested by the students.

Three-digit addition

Show this number sentence: **734 + 235 =** ______

How would you solve this problem? (734 add 200 equals 934, add 30 equals 964, add 5 equals 969)

Discuss the range of strategies that students provide, for example:

Jump 734 + 200 + 30 + 5

Split 700 + 200 + 30 + 30 + 4 + 5

Repeat the activity for these number sentences: **674 + 315 =** ______ **535 + 409 +** ______

Encourage students to discriminate in their use of calculation strategies by considering the numbers concerned. A mental strategy may often be more efficient than a formal written algorithm.

Four-digit addition

Show this number sentence: **6200 + 2385 =** ______

What would be an estimate for this number sentence? (You can round 2385 to 2300. Add 6200 equals 8500. You just look at the front end. Add the thousands as they are the biggest. So 8000.)

Discuss the students' estimation strategies, such as rounding or the 'front end' strategy.

Repeat the activity for these number sentences: **3674 + 4310 =** ______ **5535 + 4250 =** ____

Note that by being exposed to a variety of methods, students can develop the strategies that are most efficient for them.

Three-digit take away

Show this number sentence: **872 – 230 =** ______

How would you solve this problem? (872 take away 200, that is 672 take away 30 . . . 642.)

Discuss the students' strategies. Show the take-away strategy:

Take away 872 – 230 = 872 – 200 – 30 = 642

Repeat the activity for these number sentences: **682 – 160 =** ______ **767 – 220 =** ______

Three-digit counting on

Show this number sentence: **606 – 545 =** ______

How would you solve this problem? (The numbers are about the same size. You can count on from the smaller number to the larger number. 545 add 50 equals 595, add 10 equals 605, add 1 is 606, that is a difference of 61.)

Show the counting on strategy:

Counting on 545 + 50 + 10 + 1 = 606

Repeat the activity for these number sentences: **625 – 554 =** ______ **905 – 822 =** ______

Four-digit subtraction

Show this number sentence: **4569 – 3200 =** ______

What would be an estimate for this number sentence? (46 take away 32 is 14, so 46 hundred take away 32 hundred is 14 hundred, that is about 4600 take away 3200 is 1400.)

Discuss the students' strategies, including take away or counting on.

Repeat the activity for these number sentences: **4782 – 2007 =** ______ **5767 – 3998 =** ______

Changing order

Show these number sentences:

Can you see a quick way to add these? (92 + 8 is 100, 60 + 440 is 500, so 100 plus 500 is 600 ... 92 + 8 + 60 + 440 = 600.)

Encourage the students to search for compatible numbers, adding them in any order.

92 + 8 + 60 + 440 = ______

1500 + 401 + 500 + 19 = ______

9 + 7900 + 71 + 20 = ______

210 + 6 + 894 + 190 = ______

450 + 12 000 + 550 + 50 = ______

Variation

Ask the students to write some similar addition sentences of their own. Practise examples with three, four or five addends.

Comparing numbers

Show these numbers: **5200, 20 900, 39, 461, 104 610, 1000**

How does 10 461 compare to the rest of these numbers? (5200 is about half of 10 461. 104 610 is 10 times greater than 10 461. If you take away 10 000 you get 39.)

Discuss the students' responses. Emphasise the value of estimation in daily life and in checking solutions to problems. Practise similar number comparisons often to develop the students' estimation skills.

Front-end estimation

Materials

- a class set of calculators

Show this number sentence: **14 217 + 5246 + 6099 + 3482 =** ______

How would you estimate the answer to this number sentence? (You just add up the thousands digits. It is 14 plus 5 plus 6 plus 3 that is 28 000.)

Discuss the students' estimation strategies. Encourage the students to then calculate the answer. It is appropriate to use a calculator for these types of examples.

Do you think your calculation is reasonable or not? Why? (The estimate was 28 000, and the answer is 29 044. That looks correct.)

Repeat the activity for other combinations of 3-, 4- and 5-digit numbers.

Rounding off

Show these number sentences: **5473 – 2297 =** ______ **3456 + 2318 =** ______

How would you estimate the answer to these number sentences? (You can round off the hundreds. It is 54 hundred minus 22 hundred, that is 32 hundred.)

Discuss the students' estimation strategies. Encourage students to then calculate the answer.

Repeat the activity for these number sentences:

3505 – 1482 = ______ **4022 + 3855 =** ______

Three addend estimates

Show this number sentence: **3930 + 2205 + 4711 =** ______

How would you estimate the answer to this number sentence? (That is about 4000 add 2000 add 5000, about 11 000 altogether.)

Discuss the students' estimation strategies. Encourage students to then calculate the answer. Mental, written or calculator calculation may be used.

Do you think your calculation is reasonable or not? Why? (I calculated 10 846. It is very close to the estimate, so I think I am right.)

Repeat the activity for these number sentences:

4176 + 7288 + 1812 = ______ **9990 + 1437 + 5221 =** ______

Variation

Repeat the activity with four addends. Include 3-, 4- and 5-digit numbers.

Service problems

Joanne's family car is scheduled for a service at 30 000 km. The odometer currently shows 25 455 km. How many more kilometres does the car need to travel before its service?

Discuss the different strategies used by students. Invite students to justify their solutions by explaining their mathematical thinking.

Money problems

Show these amounts: **$3679** **$5187** **$8252** **$5538**

Can you make up an addition problem using these amounts?

Discuss a possible problem with the students.

How would you estimate the answer to this problem? (You use the front ends. Add 3, 5, 8 and 5 thousand, so 21 thousand.)

Discuss the students' estimation strategies. Encourage the students to then calculate the answer.

On the level addition

Show this number sentence: **3900 + 5100 = ______**

What is the answer to this number sentence? Can you tell me how you worked it out?

Discuss the different strategies used by students. Focus on the levelling strategy, for example, 3900 + 5100 = 3900 + 100 + 5100 – 100 = 4000 + 5000 = 9000.

Repeat the activity for these number sentences:

2800 + 3200 = ______ **4950 + 5050 = ______** **2250 + 6750 = ______**

On the level subtraction

Show this number sentence: **5100 – 1900 = ______**

What is the answer to this number sentence? Can you tell me how you worked it out?

Discuss the different strategies used by students. Focus on the levelling strategy, for example: 5100 – 1900 = 5100 + 100 – 1900 + 100 = 5200 – 2000 = 3200.

Repeat the activity for these number sentences:

6200 – 2800 = ______ **7400 – 3600 = ______** **6700 – 2300 = ______**

Compensation addition

Show this number sentence: **4350 + 1800 = ______**

What is the answer to this number sentence? Can you tell me how you worked it out? (You add 200 to 1800 to make it 2000, 4350 add 2000 is 6350, take away 200 gives 6150.)

Discuss the students' responses. Focus on the compensation strategy to obtain the solution.

Repeat the activity for these number sentences: **3840 + 2900 = ______** **5400 + 3950 = ______**

Compensation subtraction

Show this number sentence: **3410 – 1900 = ______**

What is the answer to this number sentence? Can you tell me how you worked it out? (3410 take away 2000 is 1410. Add 100 back on equals 1510.)

Discuss the students' responses. Focus on the compensation strategy to obtain the solution.

Repeat the activity for these number sentences: **6430 – 1800 = ______** **8760 – 2950 = ______**

Second term missing

Show this number sentence: **1503 – ______ = 799**

What does this number sentence say? (I think it says something subtracted from 1503 gives an answer of 799. I think it says you add something to 799 to get 1503.)

With aid of a number line, discuss the students' solution strategies.

Repeat the activity for these number sentences:

1435 – ______ = 626 **2672 – ______ = 1456**

Second addition

Show this number sentence: **1436 + ______ = 1642**

What does this number sentence say? Can you tell me how you can work this out? (I think it says something added to 1436 gives an answer of 1642. You can count on from 1436, add 200 that is 1636, add 6 to get 1642, the answer is 206.)

With aid of a number line, discuss the strategy of counting on. Repeat the activity for these number sentences: **1586 + ______ = 1890** **1473 + ______ = 1601**

First term missing

Show this number sentence: ______ **+ 4400 = 6000**

What does this number sentence say? (It means something added to 4400 equals 6000.) Can you tell me how you can work it out?

Discuss the different strategies used by the students. Highlight the fact that counting on is a useful strategy when the two numbers are of similar sizes. Repeat the activity for these number sentences: ______ **+ 1750 = 3000** ______ **+ 7700 = 10 000**

Find the difference

Show these numbers: **21 500 and 32 700**

How could you find out the difference between 21 500 and 32 700? (You can jump 500 up to 22 000, 10 000 up to 32 000 and then another 700 to 32 700.)

Discuss the students' responses. If necessary, use a number line to help with visualisation and mark the different jumps that could be made. Repeat the activity for the difference between:

33 200 and 43 100 **43 750 and 52 500** **61 900 and 77 400**

Find the number

Show this number: **5500**

This number is the difference between two numbers. Can you tell me what the two numbers are? (10 500 subtract 5000 equals 5500. 105 500 take away 100 000 gives an answer of 5500.)

Discuss the students' responses. Repeat the activity for other 3-, 4- and 5-digit numbers. Calculators may be used for extra motivation in these types of activities.

Multiplication and division

Year 5

The top 10

Show these numbers: **6, 42, 7**

How many different multiplication and division facts can you tell me about these numbers? (7 multiplied by 6 is 42, 6 sevens are 42, $42 \div 7 = 6$ and $42 \div 6 = 7$, $\frac{1}{7}$ of 42 is 6 and $\frac{1}{6}$ of $42 = 7$)

Revise the most challenging multiplication and division facts based around:
6×6 6×7 6×8 6×9 7×7 7×8 7×9 8×8 8×9 9×9

Multiply two by one

Show these multiplication number sentences: **15 × 8 =** ______; **23 × 4 =** ______

Can you tell me how to work out these problems? How did you work them out? (That is 4 twenties add 3 fours, that is 80 plus 12 is 92. 15×8 is three lots of 5×8, so three forties is 120. 15×8 is 5×3 multiplied by 4×2, that is 10×12 is 120.)

Allow students to work in pairs and discuss their strategies, and then present them to the rest of the class.

Halve and double

Show these number sentences: **14 × 5 =**______ **42 × 5 =** ______

Can you tell me how to work out these problems? (To multiply by 5, you multiply by 10 and then halve it . . . 14×10 equals 140, halve it equals 70. You can halve 42 and double the 5, 42×5 is the same as 21×10, that is 210.)

Allow students to work in pairs and discuss their strategies, and then present them to the rest of the class.

Open multiplication

Show this open number sentence: ______ **×** ______ **= 120**

Can you think of some numbers that complete this number sentence? How did you work it out? (4×3 equals 12 so 4×30 equals 120. 4×40 is 120 so 8×15 is also 120)

Discuss the students' strategies and encourage a variety of solutions.

Repeat the activity for these number sentences: ______ **×** ______ **= 160**, ______ **×** ______ **= 240**.

Variation

Extend the activity by using 4-digit numbers when students are ready.

Picture this

Can you imagine the number of legs on six horses and six riders? The number of wheels on 14 cars? The number of wheels on 13 cars and four motor bikes? The number of legs on nine spiders? The number of wheels on four semi-trailers?

Discuss the students' solutions. Note that the last question is open – answers could vary. Ask the students to create similar problems of their own.

Multiply by five

Show these multiplication number sentences: **32 × 5 =** ____ **52 × 5 =** ____ **72 × 5 =** ____

Can you tell me how to work out these problems? (You can multiply by 10 and halve your answer. Half of 52 is 26 so the answer is 260.)

Discuss the students' strategies.

Variation

Repeat the activity for these number sentences:

34 × 5   54 × 5   74 × 5   36 × 5   56 × 5   76 × 5   38 × 5   58 × 5   78 × 5

Nines by 5

Show these multiplication number sentences: **19 × 5 =** ____ **29 × 5 =** ____ **39 × 5 =** ____

Can you tell me how to work out these problems? (You can round off to the nearest 10 and then subtract one group of 5. That is 30 × 5 subtract 5 . . . 145.)

Discuss the students' strategies.

Variation

Repeat this activity for these number sentences:

49 × 5   59 × 5   69 × 5   79 × 5   89 × 5   99 × 5

Subtractive multiplication

Show this multiplication number sentence: **48 × 6**

Can you tell me how to work out this problem? (That is 6 × 40 plus 6 × 8, 240 plus 48 is 288. There is an easier way, you say 50 × 6 and take away 2 × 6 . . .that is 300 take away 12 equals 288.)

Discuss the students' strategies.

Variation

Repeat this activity for these number sentences:

58 × 6 = ______ **98 × 6 =** ______ **69 × 6 =** ______

Choose your numbers

Renee delivers advertising catalogues after school. The number she delivers each day is always the same 2-digit number. The number of days she works each fortnight is a 1-digit number. How many catalogues might she deliver in any one fortnight?

Ask the students to work in pairs as they investigate this problem. After sufficient time has elapsed, ask students to present their solutions. Discuss students' strategies.

(Say she delivers 68 catalogues on 5 days. That's 68 × 5. 68 × 5 is 70 × 5 – 2 × 5. That's 350 – 10 . . . 340.)

Ask students to solve the problem using another set of numbers.

Three-digit doubling

Show this number:

346

What is this number doubled?

(Double 300 is 600. Double 40 is 80. Double 6 is 12. That's 692 altogether. Double 350 is 700. Take away 2 × 4 leaves 692.)

Call for answers. Discuss students' strategies.

Repeat the activity for:

159 **267** **471**

Opposite facts

Show these division facts:

45 ÷ 5 = ______ **56 ÷ 7 =** ______ **63 ÷ 9 =** ______

How can you use multiplication to answer these questions? (9 fives are 45, so 45 ÷ 5 equals 9. There are 8 groups of 7 in 56 . . . If you divide 56 into groups of 7, there are 8 of them.)

Discuss the students' responses. Repeat the activity to revise the most challenging multiplication and division facts based around:

6 × 6 6 × 7 6 × 8 6 × 9 7 × 7 7 × 8 7 × 9 8 × 8 8 × 9 9 × 9

Divide two by one

Show these division number sentences: **64 ÷ 4 =** ______ **90 ÷ 6 =** ______

Can you tell me how to work out this problem? (That is 60 divided by 4, add on 4 divided by 4, so 15 add 1 is 16. Dividing by 4 is the same as dividing by 2 twice, so half of 64 is 32, then halve it again and you get 16. 90 ÷ 6 is the same as 90 ÷ 3 ÷ 2, that is 30 ÷ 2, equals 15.)

Ask the students to work in pairs to discuss their strategies and present their strategies to the rest of the class.

Divide by five

Show these division number sentences:

170 ÷ 5 = ______ **280 ÷ 5 = ______** **360 ÷ 5 = ______**

Can you tell me how to work out this problem? (You can divide by 10 and double your answer, 280 divided by 10 is 28, double 28 is 56.)

Ask the students to work in pairs to discuss their strategies and present their strategies to the rest of the class.

Variation

Repeat the activity for these division number sentences:

190 ÷ 5 = ______ **390 ÷ 5 = ______** **480 ÷ 5 = ______**

Three-digit halving

Show this number:

What is this number halved?

Call for answers. Discuss students' strategies.

(You halve the hundreds, tens and halve the units. Add them together … That's 300 + 20 + 1. Half of 640 is 320. Half of 2 is 1. That makes 321.)

Repeat the activity for:

Year 6

What is the number?

Materials

- counters
- multiplication grid

Show a blank multiplication grid:

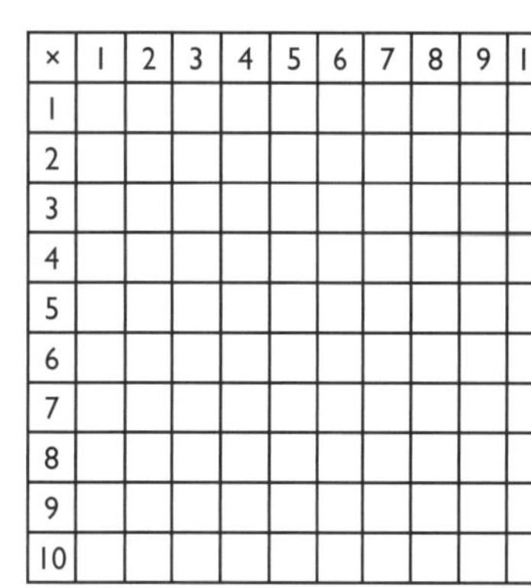

×	1	2	3	4	5	6	7	8	9	10
1										
2										
3										
4										
5										
6										
7										
8										
9										
10										

Can you imagine the missing numbers?

Place a counter on any square.

What is this number on the multiplication chart? How do you know?

Repeat this activity often to revise and consolidate multiplication basic facts. Concentrate on the 10 most challenging facts:

6 × 6 **6 × 7** **6 × 8** **6 × 9** **7 × 7**

7 × 9 **8 × 8** **8 × 9** **9 × 9**

Product puzzles

Show this 3 × 3 grid:

1	3	
5	7	

What will happen if we multiply the numbers in the rows and columns of this grid?

Discuss the students' responses. Multiply the numbers horizontally and vertically, showing the products in the squares provided. Then multiply the products to get the number bottom right square. The answer should be the same, horizontally and vertically.

1	3	3
5	7	35
5	21	105

Does this happen for combinations of all 1-digit numbers?

Encourage the students to investigate single-digit product puzzles of their own.

Variation

Students can use a calculator to explore similar products with 2-digit numbers.

Mental workout

Ask the students to imagine the number 15. Tell the students that they must do some progressive mental calculations to get to an answer.

The starting number is 15 . . . add 5 . . . take away 6 . . . divide by 2 . . . multiply by 8 . . . add 4 . . . divide by 4 . . . What is the answer?

Repeat for other random sequences. Ask the students to prepare some sequences of their own and invite them to share their sequences with the class.

Multiply two by one

Show these multiplication number sentences: **18 × 8 =** _______ **32 × 4 =** _______

Can you tell me how to work out these problems? (18 × 8 is 6 × 3 multiplied by 4 × 2, that is 12 × 12 = 144. That is 4 thirties add 4 twos, that is 120 plus 8 is 128.)

Ask the students to work in pairs and discuss their strategies, then present them to the rest of the class. Encourage the students to choose their strategies by considering the numbers first – a mental strategy may often be more efficient than a formal written algorithm.

Multiply three by one

Show these multiplication number sentences: **132 × 6 =** _______ **240 × 7 =** _______

Can you tell me how to work out these problems? (That is 600 plus 180 plus 12 = 792. Add 200 × 7 and 40 × 7 = 1400 plus 280 equals 1680.)

Ask the students to work in pairs and discuss their strategies, and then present them to the rest of the class. Encourage the students to choose their strategies by considering the numbers first – a mental strategy may often be more efficient than a formal written algorithm.

Multiply two by two

Show these multiplication number sentences: **60 × 15 =** ______ **21 × 13 =** ______

Can you tell me how to work out these problems? (That is 60 × 5 × 3, that is 300 × 3 that equals 900. You work out 20 × 13 plus 1 × 13, that is, 260 add 13 is 273.)

Ask the students to work in pairs and discuss their strategies, then present them to the rest of the class. Encourage the students to choose their strategies by considering the numbers first – a mental strategy may often be more efficient than a formal written algorithm.

Halve and double

Show these number sentences: **280 × 5 =** ______ **462 × 5 =** ______

Can you tell me how to work out these problems? (280 × 10 equals 2800 . . . halve it, that equals 1400. You can halve 462 and double the 5, that is 231 × 10, that is 2310.)

Ask the students to work in pairs and discuss their strategies, and then present them to the rest of the class. Encourage the students to choose their strategies by considering the numbers first – a mental strategy may often be more efficient than a formal written algorithm.

Multiply by five

Show these number sentences:

41 × 5 = ______ **61 × 5 =** ______ **81 × 5 =** ______

Can you tell me how to work out these problems? (You can multiply by 10 and halve your answer, half of 810 is 405. 5 sixties are 300, 5 ones are 5, the answer is 305.)

Ask the students to work in pairs and discuss their strategies, and then present them to the rest of the class. Encourage the students to choose their strategies by considering the numbers first – a mental strategy may often be more efficient than a formal written algorithm.

Variation

Practise multiplying 2-digit numbers ending in 3, 5 and 7 by 5.

Nines by five

Show these multiplication number sentences:

69 × 5 = ______ **99 × 5 =** ______ **299 × 5 =** ______

Can you tell me how to work out these problems? (You can round the number off to the nearest 10 or 100, then you multiply it by 5 and subtract one group of 5. Or 99 × 5, that is 100 × 5, subtract 5 you get 495.)

Ask the students to work in pairs and discuss their strategies, and then present them to the rest of the class. Encourage the students to choose their strategies by considering the numbers first – a mental strategy may often be more efficient than a formal written algorithm.

Variation

Repeat the activity for other 3-digit numbers ending in 99 or 98.

Open multiplication

Show these open number sentences:

______ × ______ = 1500 **______ × ______ = 2400**

Can you think of some numbers that complete these number sentences? (There are 24 hundreds in 2400, so it is 24 × 100. Three fives are 15, so 3 × 500 is 1500.)

Discuss the students' strategies. Encourage a variety of solutions.

Repeat the activity for: **______ × ______ = 1750** **______ × ______ = 3250**

Variation

Extend the activity by using examples with 3 multiplicands.

Mixed sentences

Show this open number sentence: **______ × ______ + ______ = 1000**

Can you think of some numbers that complete this number sentence? (You can start by adding 100, then there is 900 left. 3 × 300 equals 900. Or 10 × 40 is 400, and add another 600 to get 1000.)

Discuss the students' strategies. Encourage a variety of solutions.

Extension

Use division and subtraction symbols in the open number sentences.

Multiplication patterns

Show these multiplication number sentences:

4 × 16 = ______ **4 × 160 = ______** **4 × 1600 = ______** **4 × 16 000 = ______**

Can you tell me how to work out these problems? What patterns do you notice? (It is 4 groups of 16 each time. The answer multiplies by 10 each time. Four groups of 16 equals 64. Four groups of 16 tens is 64 tens. Four groups of 16 hundreds is 64 hundreds.)

Discuss the students' strategies. Ask the students to devise some multiplication patterns of their own. Calculators may be used.

Subtractive multiplication

Show these multiplication number sentences: **78 × 6 = ______** **19 × 13 = ______**

Can you tell me how to work out these problems? Which strategies did you use? (That is 80 × 6 take away 2 × 6, that is 480 − 12 = 468. I would multiply 20 × 13 and take away 1 × 13, that is 260 − 13 = 247.)

Discuss the students' strategies. Repeat the activity for these number sentences:

98 × 6 = ______ **29 × 14 = ______**

Estimate and multiply

Materials

- a class set of calculators

Show these number sentences:

327 × 4 = ______ **4212 × 3 = ______** **10 523 × 8 = ______**

How would you estimate the answer to these number sentences? (You concentrate on the front end – the thousands digits, it is 4 thousand times 3, about 12 000. You can round off to the nearest hundred, say 10 500 × 8, that is 80 000 plus 8 × 500 = 84 000.)

Discuss the students' estimation strategies. Encourage the students to then calculate the answer. It is appropriate to use a calculator for this activity.

Do you think your calculation is reasonable or not? Why? (The estimate was 12 000 and the answer is 12 636. I would say itis correct.)

Repeat the activity for other 3-, 4- and 5-digit numbers.

Multiply by 20

Show these multiplication number sentences:

23 × 20 = ______ **47 × 20 = ______** **125 × 20 = ______**

Can you tell me how to work out these problems? Which strategies did you use? (To multiply by 20, you multiply by ten first and then double your answer. 125 × 10 is 1250, double it is 2500.)

Discuss the students' strategies.

Variation

Repeat the activity multiplying other 2- and 3-digit numbers by 20.

Multiply by 100 (1)

Materials

- a place value chart

Show these number sentences:

83 × 100 = ______ **431 × 100 = ______** **3256 × 100 = ______**

Can you tell me how to work out these problems?

Discuss students' responses.

Why do we add 2 zeros when we multiply by 100?

With the aid of the place value chart explain that multiplying by one hundred shifts everything two places to the left.

× 100 ⟵ 2 places

Hundred Thousands	Ten Thousands	Thousands	Hundreds	Tens	Ones
		3	2	5	6
3	2	5	6	0	0

The number becomes larger by two powers of ten and two zeros are inserted as place holders.

Variation

Practise multiplying numbers up to 6 digits by 100.

Multiply by 50

Show these multiplication number sentences:

83 × 50 = ______ **91 × 50 = ______** **79 × 50 = ______**

Can you tell me how to work out these problems? (If it starts with a low number, such as 1, 2 or 3, or ends with a high number, such as 8 or 9, you can round it off to the nearest 10, multiply by 100, then halve it and subtract 50. To multiply by 50, you can multiply by 100 and then halve it, that is 4150. For 79 × 50, round up to 80, then multiply by 100 to get 8000, halve that to get 4000 and subtract 50 to get 3950.)

Discuss the students' strategies.

Variation

Practise multiplying other 2-digit odd numbers by 50.

Product game

Materials

- two sets of number cards from 0 to 9
- a calculator

Show an open number sentence:

______ × ______ = ______ (Team 1) ______ × ______ = ______ (Team 2)

Explain the rules for the product game.

The product game rules

1. Divide the class into two teams and invite four students from each team to choose a number card.
2. Students can put their cards in any position on the open number sentence except the answer.
3. When all the students have placed their cards, students can estimate the answer and the team that makes the highest product possible wins the game.

Discuss the students' estimation strategies.

Variation

Repeat the game with the aim of making the lowest possible quotient.

Extension

Repeat the activity using a grid with the written algorithm:

Cross quotients

Show this 3 × 3 grid:

48	8	
6	2	

What will happen if we divide the numbers at the top of a column or the left of a row by the number below or next to them?

Discuss the students' responses. Divide the numbers horizontally and vertically, writing quotients in the squares provided. Then divide the quotients. The answer in the bottom right square should be the same, horizontally and vertically.

48	8	6
6	2	3
8	4	2

Does this happen for other basic fact combinations?

Encourage the students to investigate other quotient puzzles of their own.

Divide three by one

Show these division number sentences: **186 ÷ 6 =** ______ **294 ÷ 7 =** ______

Can you tell me how to work out these problems? (180 divided by 6 is 30, 6 divided by 6 is 1, so that is 31. 7 forties are 280, 7 twos are 14, 40 plus 2 equals 42.)

Ask the students to work in pairs to discuss their strategies and present their strategies to the rest of the class.

Divide two by two

Show these division number sentences: **90 ÷ 15 =** ______ **75 ÷ 15 =** ______

Can you tell me how to work out these problems? (75 divided by 15. Here you can divide 60 by 15 and then divide 15 by 15, that is 4 plus 1 equals 5. That is 90 divided by 3 divided by 5, 30 divided by 5 is 6.)

Ask the students to work in pairs to discuss their strategies and present their strategies to the rest of the class

FRACTIONS AND DECIMALS

Year 5

Fraction meanings

Show the fraction: $\frac{2}{3}$

What does $\frac{2}{3}$ mean? ($\frac{2}{3}$ means 2 ÷ 3) How many ways can you show me the idea of $\frac{2}{3}$?

Discuss the students' responses. Encourage the students to make up stories or devise number situations involving the idea of $\frac{2}{3}$.

(There were 3 cupcakes, 1 had icing and 2 didn't. The pancake was divided into 3 equal parts and my brother ate 2 parts out of 3. I ate $\frac{1}{3}$ of one pizza and of $\frac{1}{3}$ another pizza. We shared 2 pizzas equally between 3 people. Our car is $\frac{2}{3}$ as long as our boat.)

Taking the cake

Two children were offered a fraction of a birthday cake. They could take either $\frac{1}{6}$ or $\frac{1}{3}$. Which fraction did they take and why? (If the cake is divided among 6 people the slices are smaller than if it is divided among 3 people. I would take $\frac{1}{3}$ because it would be bigger, and I like cake. I am not hungry, so I will have $\frac{1}{6}$ because it is smaller.)

Discuss the different strategies used by students. Encourage the students to justify their answers, using diagrams or concrete materials to help if necessary.

Variation

Repeat the activity for other fractions such as $\frac{1}{2}$, $\frac{1}{4}$ and $\frac{1}{3}$.

Paper strips

Materials

- paper strips
- a pen

Fold a paper strip into thirds. Unfold the strip and mark the fold lines. Show the strip to the students.

What fractions do you see?

Discuss the idea of one-third as one equal part of one whole that has been divided into three equal parts. Fold each of the thirds in half.

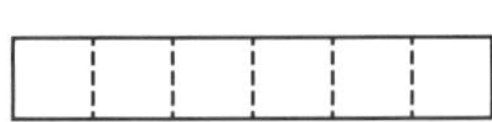

What fractions do you see now? What can you tell me about these thirds and sixths? (There are six parts, they are sixths. Thirds are larger than sixths. Sixths are smaller than thirds. Two-sixths is the same as one-third. Six-sixths is the same as three-thirds. If you take half of a third you get one-sixth.)

Distribute the paper strips and allow the students to investigate the ideas of equivalence for themselves.

Fraction bars

Show these three fraction bars:

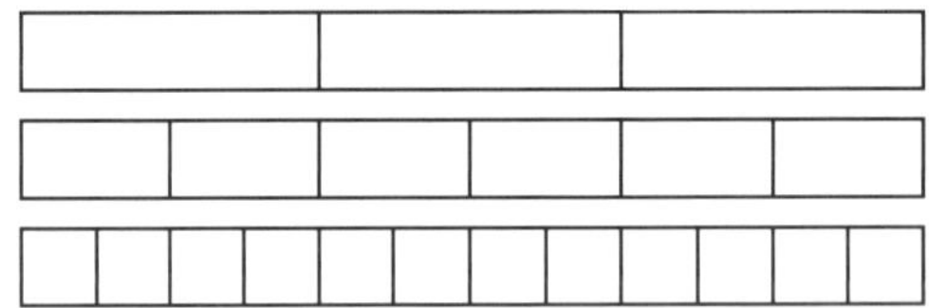

Ask the students to shade $\frac{1}{3}$ of the upper bar, $\frac{2}{6}$ of the middle bar and $\frac{4}{12}$ of the lower bar. Label the bars as $\frac{1}{3}$, $\frac{2}{6}$ and $\frac{4}{12}$.

What do you notice about $\frac{1}{3}$, $\frac{2}{6}$ and $\frac{4}{12}$? (They are the same length. They take up the same space.)

Discuss the students' responses. Introduce the term 'equivalent fractions' using these fraction bars as examples.

Equivalent strips

Materials

- paper strips
- pins or Blu-Tack
- a pen

Fold a paper strip into thirds, another into sixths and another into twelfths. Show the fraction strips to the class. Label the ends of each strip as 0 and 1. Ask the students to help you label each interval on the fraction strips between 0 and 1.

0 $\frac{1}{3}$ $\frac{2}{3}$ 1

0 $\frac{1}{6}$ $\frac{2}{6}$ $\frac{3}{6}$ $\frac{4}{6}$ $\frac{5}{6}$ 1

0 $\frac{1}{12}$ $\frac{2}{12}$ $\frac{3}{12}$ $\frac{4}{12}$ $\frac{5}{12}$ $\frac{6}{12}$ $\frac{7}{12}$ $\frac{8}{12}$ $\frac{9}{12}$ $\frac{10}{12}$ $\frac{11}{12}$ 1

How many different equivalent fractions can you see on these strips?

Discuss the students' responses in terms of equivalent fractions.

Equivalent fractions

Show these fractions: $\frac{1}{3}$ $\frac{2}{6}$ $\frac{4}{12}$

Can you see a pattern or relationship between these three fractions? (If you double the top and bottom of $\frac{1}{3}$ you get $\frac{2}{6}$. If you divide 6 by 2 you get 3, and 12 divided by 3 is 4, that is 2 quarters.)

Discuss the students' responses. Refer to fraction bars or fraction strips to review equivalent fractions. Ask the students to suggest examples of other equivalent fractions.

Sixths line up

Materials

- a 1-m length of string
- pegs

a set of fraction cards labelled $\frac{1}{6}, \frac{3}{6}, \frac{4}{6}, \frac{6}{6}, \frac{7}{6}, \frac{9}{6}, \frac{10}{6}, \frac{11}{6}$.

Tell the students that they are going to construct a number line segment between 0 and 2.

Invite a student to select a fraction card.

Can you estimate where to place this numeral card on the number line? Why did you place it there? ($\frac{11}{6}$ is almost as large as 2, so it goes near the end of the line.)

Ask the students to justify their placements. Repeat the activity, attaching the cards in the positions suggested by the students. Count in sixths forwards and backwards along the number line.

Variation

Repeat the activity with different fraction cards, for example,

$\frac{1}{12}, \frac{3}{12}, \frac{4}{12}, \frac{9}{12}, 1\frac{2}{12}, 1\frac{3}{12}, 1\frac{5}{12}, 1\frac{11}{12}$.

Unit fraction line up

Materials

- a 1-m length of string
- pegs

a set of fraction cards labelled $\frac{1}{2}, \frac{1}{3}, \frac{1}{4}, \frac{1}{5}, \frac{1}{6}, \frac{1}{8}, \frac{1}{10}, \frac{1}{12}$

Tell the students that they are going to construct a number line segment between 0 and 1.

Invite a student to select a fraction card.

Can you estimate where to place this numeral card on the number line? Why did you place it there? (I think $\frac{1}{3}$ goes there because it is bigger than $\frac{1}{5}$ but smaller than $\frac{1}{2}$.)

Ask the students to justify their placements. Repeat the activity, attaching the cards in the positions suggested by the students.

Mixed line up

Materials

- a 1-m length of string
- pegs
- a set of fraction cards labelled $\frac{1}{6}, \frac{2}{3}, \frac{3}{4}, \frac{4}{5}, \frac{1}{2}, \frac{5}{12}$

Tell the students that they are going to construct a number line segment between 0 and 2.

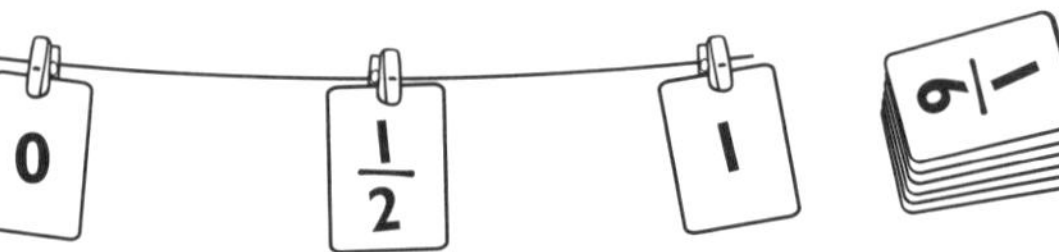

Invite a student to select a card.

Can you estimate where to place this numeral card on the number line? Why did you place it there? ($\frac{2}{3}$ is smaller than $\frac{3}{4}$, but it is bigger than $\frac{1}{6}$.)

Ask the students to justify their placements. Repeat the activity, attaching the cards in the positions suggested by the students.

Variation

Repeat the activity with other selections of mixed fractions.

Around one whole

Seven children were cooking pancakes. Each received $\frac{1}{6}$ of a pancake. Altogether did they have more than one pancake or less than one pancake? (7 sixths, that is 6 sixths – one whole – and one more sixth. I think they had more than one, because $\frac{7}{6}$ is more than one.) Exactly how many pancakes did they have?

Encourage the students to discuss their answers in pairs and explain their answers using diagrams or concrete materials. Encourage the students to record the answer as an improper fraction, $\frac{7}{6}$, and as a mixed numeral, $1\frac{1}{6}$.

Mixed numerals

Show the mixed numeral: $1\frac{3}{4}$

What do we call this type of number? How many quarters are there in $1\frac{3}{4}$? (There are 4 quarters in 1 and another 3 quarters, that is 7 quarters.) How could you prove your answer?

Discuss the students' responses. Encourage the students to use diagrams to explain their thinking. The difference between improper fractions and mixed numbers can also be contrasted in this activity.

Variation

Repeat the activity for: $1\frac{2}{3}$, $2\frac{3}{4}$, $3\frac{5}{6}$.

Lamington bars

Jasmin baked some lamingtons for the family. If she ate $\frac{1}{3}$ of a lamington, what fraction was left? How could you prove your answer?

Ask the students to discuss their solutions and explain their answers, using diagrams or concrete materials. Introduce notation to represent the problem as: $1 - \frac{1}{3} = \frac{2}{3}$.

Pancakes

Tony cooked pancakes for his family. They ate 1 pancake each and his father ate an extra $\frac{1}{2}$ a pancake.

How many pancakes did Tony's father eat? How could you prove your answer?

Ask the students to discuss their solutions and explain their answers, using diagrams or concrete materials. Introduce notation to represent the problem as: $1 + \frac{1}{2} = 1\frac{1}{2}$.

Pizza thief

Anna cooked 4 pizzas and left them to cool. Her sister was very hungry and ate $\frac{1}{3}$ of one pizza. How many pizzas remained? How could you prove your answer?

Ask the students to discuss their solutions and explain their answers, using diagrams or concrete materials. Introduce notation to represent the problem as: $4 - \frac{1}{3} = 3\frac{2}{3}$.

Lamington total

Jordan and Nicholas bought one lamington each. Jordan ate $\frac{5}{6}$ of his lamington and Nicholas ate $\frac{3}{6}$ of his lamington. How many lamingtons did they eat altogether? How could you prove your answer?

Ask the students to discuss their solutions and explain their answers, using diagrams or concrete materials. Introduce notation to represent the problem as: $\frac{5}{6}+\frac{3}{6}=\frac{8}{6}=1\frac{2}{6}=1\frac{1}{3}$.

Pizza leftovers

Bianca collected the leftover pizzas and put them in the fridge. She collected one whole pizza, $\frac{3}{4}$ of pizza, $\frac{1}{4}$ of a pizza and $\frac{2}{4}$ of another pizza. How much pizza did she collect altogether? How could you prove your answer?

Ask the students to discuss their solutions and explain their answers, using diagrams or concrete materials. Introduce notation to represent the problem as: $1+\frac{3}{4}+\frac{1}{4}+\frac{2}{4}=2\frac{2}{4}=2\frac{1}{2}$.

Chocolate bar

Melissa bought a chocolate bar. She divided it into sixths. She took $\frac{1}{6}$ for herself and gave $\frac{1}{6}$ to her friend Brianna. What fraction of the original chocolate bar did the girls now have? What fraction of the chocolate bar was left?

Ask the students to discuss their solutions and explain their answers, using diagrams or concrete materials. Introduce notation to represent the problem as: $1-\frac{1}{6}-\frac{1}{6}=\frac{4}{6}$.

Investigating hundredths

Materials

- a Base 10 hundred block

Show the students a Base 10 hundred block.

If this block represents one whole number, what would one small square represent? ($\frac{1}{100}$, 0.01, there is one hundredth, there are no tenths.)

Discuss the students' responses.

If we had 35 squares of the hundred block, what would this look like as a fraction or decimal number? (35 hundredths, $\frac{35}{100}$, 0.35) How can we relate hundredths to our money system? (There are 100c in \$1.00 so 1 cent is $\frac{1}{100}$. 1 cent can also be written as \$0.01.)

Variation

Show these money amount as fractions:

3 dollars and 6 cents   8 dollars and 40 cents   82 cents

Show these fractions as money:

1whole $\frac{24}{100}$ 4 wholes $\frac{5}{100}$.

Imagining thousandths

Materials

- Base 10 materials

Show the students a Base 10 cube and a hundred (flat).

If the large cube represents one whole, what fraction is the 'flat'? (There are 10 flats in the cube so its $\frac{1}{10}$*.)*

Discuss the students' responses. Show the students a ten (long).

If the large cube represents one whole, what fraction is the 'long'? (It is 1 out of 100, that is 1 hundredth.)

Repeat the question for the one (short). Develop the idea of thousandths. Compare a handful of 'shorts' to the large cube. Repeat the activity to demonstrate varying amounts of thousandths.

Investigating thousandths

Materials

- Base 10 materials

Show the students 12 'shorts/ones' and the Base 10 cube.

What fraction of this large cube are these 'shorts/ones'? (It is 12 out of 1000. 12 thousandths.)

Discuss answers and develop the idea of thousandths.

What are all the different ways in which we can write 12 thousandths? (12 out of 1 thousand, 12 thousandths, $\frac{12}{1000}$*, 0.012, 1 hundredth and two thousandths.)*

Repeat the activity: compare 4 longs/tens and 1 flat/hundred, 2 longs/tens and 3 shorts/ones to a large Base 10 cube/thousand.

Decimal order (1)

Materials

- a set of decimal cards labelled 0.09, 0.45, 0.34, 0.67, 0.91

Choose five students to select a card each and show it to the class.

Can you place these decimals in order from the smallest to the largest? How do you know which decimal is the smallest or largest? (0.91 is largest because the 9 stands for 9 tenths. There are 9 tenths and one hundredth in 0.91.)

Discuss the students' responses in terms of place value of the decimals.

Decimal line up

Materials

- a 1-m length of string
- a set of number cards from 0 to 9
- pegs
- a set of decimal cards for 1.014, 1.3, 2.413, 2.4, 3.1, 3.09; 4.763, 5.7, 5.17, 8.382

Tell the students that they are going to construct a number line segment between 0 and 10 using decimals.

Invite a student to select a decimal card.

Can you estimate where to place this numeral card? Why did you place it there?

Repeat the process, attaching the cards in the positions suggested by the students. Ask the students to justify their suggestions.

Fine order

Materials

- string
- a set of number cards from 0 to 3
- pegs
- a set of decimal cards for 2.52, 2.25, 2.50, 2.05

Can you order these decimals from smallest to largest? Which is the smallest? Which is the largest? How do you know? (The ones with the 5 tenths are the largest. Then you look at the hundredths column.)

Ask the students to place the cards in ascending order on the number line. Discuss the students' suggestions in terms of place value.

Decimal order (2)

Materials

- a set of decimal cards labelled 0.132, 0.247, 0.475, 0.295, 0.467

Ask five students to select a card and show it to the class.

Can you place these decimals in order from smallest to largest? (0.132, 0.247, 0.295, 0.467, 0.475; 0.475 is largest because the 47 stands for 47 hundredths.)

Discuss the students' responses in terms of the place value of the decimals.

Close order

Materials

- a set of decimal cards labelled 0.081, 0.801, 0.810, 0.018, 0.880

Ask five students to select a card and show it to the class.

Can you order these decimals from largest to smallest? (0.880, 0.810, 0.801, 0.081, 0.018; 0.018 is smallest because the 18 stands for 18 thousandths.)

Discuss the responses in terms of the place value of the decimals.

Mixed order

Materials

- a set of decimal cards labelled 0.7, 0.72, 0.725, 0.791, 0.708

Ask five students to select a card and show it to the class.

Can you order these decimals from smallest to largest? (0.7, 0.708, 0.72, 0.725, 0.791; 0.7 is the same as 700 thousandths. That's the smallest.)

Discuss the responses in terms of the place value of the decimals. Note that it is useful to rewrite 0.7 and 0.72 to three decimal places (0.700 and 0.720) so that they may be compared more easily.

Year 6

Taking the cake

Two children were offered a fraction of a birthday cake. They could take either $\frac{1}{2}$ or $\frac{2}{5}$. Which fraction did they take and why? $\frac{2}{5}$ is less than $\frac{1}{2}$ because it is 2 parts out of 5. I would take $\frac{1}{2}$ because I am hungry.

Discuss the different strategies used by students. Encourage the students to use diagrams or concrete materials to help, if necessary. Repeat the activity for other fractions such as $\frac{1}{2}$ and $\frac{2}{3}$ or $\frac{3}{4}$ and $\frac{4}{5}$.

More than a pancake

Dad made 6 pancakes to be shared equally among 5 children. How many pancakes did they each receive? (There are enough pancakes for one whole pancake each. There is one pancake left over which can be shared among 5 children.)

Encourage the students to work in pairs to discuss their answers, using diagrams if necessary.

Record the answer using fraction notation: $6 \div 5 = 1\frac{1}{5}$.

Less than a pancake

Mum made 4 pancakes to be shared equally among 5 children. What fraction of a pancake did they each receive? (There are not enough pancakes for one whole pancake each. The answer must be less than one.)

Encourage the students to work in pairs to discuss their answers, using diagrams if necessary.

Record the answer using fraction notation: $4 \div 5 = \frac{4}{5}$.

Fraction bars

Show these fraction bars:

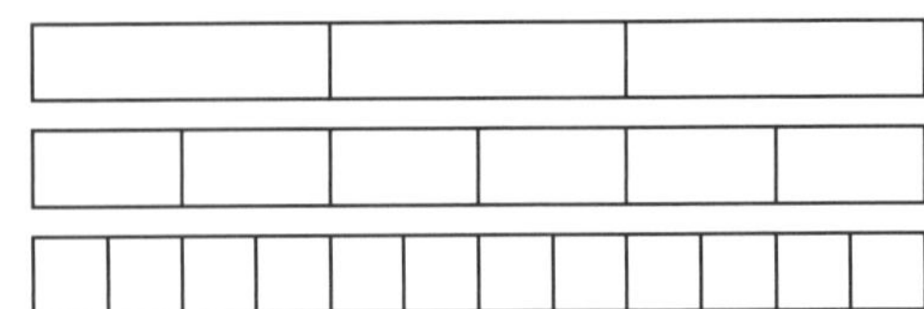

Ask the students to shade $\frac{2}{3}$ of the upper bar, $\frac{4}{6}$ of the middle bar and $\frac{8}{12}$ of the lower bar. Label the bars as $\frac{2}{3}$, $\frac{4}{6}$ and $\frac{8}{12}$.

What do you notice about $\frac{2}{3}$, $\frac{4}{16}$ and $\frac{8}{12}$?

Discuss the students' responses in terms of equivalent fractions. Explain the strategy of multiplying or dividing by a common factor to produce equivalent fractions.

Can you suggest other fractions that are equivalent to $\frac{2}{3}$?

Paper strips

Materials

- paper strips
- a pen

Ask the students to fold strips into eighths.

Which equivalent fractions can you find on your strip?

Discuss the equivalence of eighths, quarters and halves in the strip.

(One half is the same as 2 quarters or 4 eighths. 6 eighths is the same as 3 quarters.)

Variation

Repeat the activity for sixths and twelfths.

Lowest form

Show these fractions: $\frac{4}{16}$ $\frac{6}{12}$ $\frac{5}{20}$ $\frac{9}{24}$ $\frac{16}{100}$

How could we write these fractions in their lowest equivalent form? (You look for a number that will divide the numerator and denominator.)

Discuss the students' responses. Develop the idea of reducing each fraction to its lowest equivalent form by dividing the numerator and the denominator by a common factor. Use notation to represent the equivalence, for example: $\frac{4}{16} = \frac{2}{8} = \frac{1}{4}$.

Fractions line up

Materials

- a 1-m length of string
- pegs
- a set of fraction cards labelled $1\frac{1}{2}$, $2\frac{1}{4}$, $2\frac{2}{3}$, $2\frac{3}{4}$, $3\frac{1}{4}$, $4\frac{1}{3}$, $6\frac{1}{2}$, $9\frac{1}{10}$, 10 and 0.

Tell the students that they are going to construct a number line segment between 0 and 10. Invite a student to select a fraction card.

Can you estimate where to place this fraction card on the number line? Why did you place it there?

Repeat the activity, attaching the cards in the positions suggested by the students.

Variation

Repeat the activity with other selections of fraction and decimals cards using numbers greater than 1.

Jellybean shares

Three children shared a packet of 24 jellybeans. Each child received $\frac{1}{4}$ of the total number of jellybeans. (If they received $\frac{3}{4}$, $\frac{1}{4}$ must be left.) How many jellybeans did each child receive? ($\frac{1}{4}$ of 24 is the same as 24 divided by 4. That is 6, so three children got 6 each.)

How many jellybeans were left? (6) What fraction of the jellybeans was shared? What fraction of the jellybeans was left?

Discuss the students' strategies. Encourage the students to solve the problem using diagrams or concrete materials.

Lamington denominators

Laura and Emily bought one lamington each. Emily ate $\frac{2}{3}$ of her lamington and Laura ate $\frac{1}{6}$ of her lamington. How many lamingtons did they eat altogether? How could you prove your answer?

Discuss the students' strategies. Encourage the students to solve the problem using diagrams or concrete materials.

Introduce notation to represent the problem as: $\frac{2}{3}+\frac{1}{6}=\frac{4}{6}+\frac{1}{6}=\frac{5}{6}$.

Common denominator

Patrick baked a cake for his sister's birthday. His sister ate $\frac{1}{10}$ of the cake and Patrick ate $\frac{2}{5}$ of the cake. What fraction of the cake did they eat? What fraction of the cake was left? (You must change the $\frac{2}{5}$ to $\frac{4}{10}$ so they have the same denominator . . .)

Discuss the students' strategies. Encourage the students to solve the problem using diagrams or concrete materials. Introduce notation to represent the problem as:

$\frac{1}{10}+\frac{2}{5}=\frac{1}{10}+\frac{4}{10}=\frac{5}{10}$ and $1-\frac{5}{10}=\frac{5}{10}=\frac{1}{2}$.

Pizza portions

Carmela made pizzas for her three brothers. Each boy ate exactly $\frac{2}{5}$ of a pizza. Altogether, how much pizza did the boys eat? How could you prove your answer?

Discuss the students' strategies. Encourage the students to solve the problem using diagrams or concrete materials. Introduce notation to represent the problem as: $3 \times \frac{2}{5} = \frac{2}{5} + \frac{2}{5} + \frac{2}{5} = \frac{6}{5}$ and $1\frac{1}{5}$.

Chocolate fractions

Asheni bought two chocolate bars. She took $\frac{3}{4}$ of one bar for herself and gave $\frac{1}{2}$ of the other bar to her friend Bethea. How much chocolate do the girls have? How much chocolate is left?

Discuss the students' strategies. Encourage the students to solve the problem using diagrams or concrete materials. Introduce notation to represent the problem as:

$$\frac{3}{4} + \frac{1}{2} = \frac{3}{4} + \frac{2}{4} = \frac{5}{4} = 1\frac{1}{4}$$

$$2 - 1\frac{1}{4} = \frac{3}{4}$$

Pancake problem

Show this recipe for pancakes:

If this recipe serves 4 people, how could John change it to suit 8 people? What if there were 12 people?

Ask the students to discuss the problem with a partner. Discuss the students' strategies. Develop the idea of proportion in the discussion.

Unit fractions

Show these fractions: $\frac{1}{3} \times 24$ $\frac{1}{4}$ of 32 $\frac{1}{5} \times 45$ $\frac{1}{6}$ of 54

How would you calculate the answers to these questions? (3 × 8 equals 24 so $\frac{1}{3}$ of 24 is 8.)

Discuss the students' strategies. Relate the calculation of unit fractions to multiplication and division facts.

$\frac{1}{6} \times 54 = 9$ $9 \times 6 = 54$ $54 \div 6 = 9$ $\frac{1}{9} \times 54 = 6$ $6 \times 9 = 54$ $54 \div 9 = 6$

How many each?

Three children shared a packet of 12 jellybeans. One child received $\frac{1}{3}$ of the jellybeans, another received $\frac{1}{6}$ and another received $\frac{1}{12}$. How many jelly beans did they each receive? Which person received the most jelly beans? ($\frac{1}{3}$ of 12 is 4 jelly beans. $\frac{1}{12}$ of 12 is 1. The person who took $\frac{1}{3}$ got the most.)

Discuss the students' strategies. Encourage the students to solve the problem using a diagram or concrete materials to help.

Anzac biscuits

Melanie cooked a batch of 21 Anzac biscuits. She took $\frac{2}{3}$ of them to school to share with her friends. How many biscuits did she take to school?

Discuss the students' strategies. Develop the notation for the problem:

$$\frac{2}{3} \text{ of } 21 = \frac{2}{3} \times 21 = 2 \times \frac{1}{3} \times 21 = 2 \times 7 = 14$$

Repeat the activity for situations involving $\frac{3}{4}$ of 24; $\frac{5}{6}$ of 18; $\frac{2}{5}$ of 30.

Personal best

Rebecca wanted to cut 0.3 minutes off her cross-country time. How many seconds are there in 1.3 minutes?

Discuss the students' strategies, using a clock or stopwatch if necessary. Develop the notation for the problem:

$$0.3 \text{ of } 60 = \frac{3}{10} \times 60 = 3 \times \frac{1}{10} \times 60 = 3 \times 6 = 18$$

Repeat the activity for situations involving 0.4 of 30; 0.7 of 50; 0.9 of 90.

At the carnival

At the school athletics carnival, four children recorded these results.

Name	100 metre sprint	200 metre sprint	Long jump	High jump
Eddle	14.96 seconds	32.66 seconds	3.35 m	1.25 m
Suzi	15.07 seconds	32.04 seconds	3.56 m	1.34 m
Daniel	16.75 seconds	33.64 seconds	3.21 m	1.21 m
Emma	16.22 seconds	33.56 seconds	3.39 m	1.07 m

Which person was fastest over 100 m and over 200 m? Who made the best long jump and the best high jump? What was the difference between Suzi's and Emma's high jump? What was the difference between Suzi's and Daniel's long jump? Who do you think is the best athlete of the four competitors? Why?

Discuss the students' responses and ask them to explain their thinking.

Variation

Using the same table, ask the students to devise some questions of their own.

Estimating decimals

Materials

- a class set of calculators

The cost for 65 L of petrol is $68.84. Is the price per litre $1.98, 92.9c, 105.9c or 98.9c? Sausages cost $12.90 per kilogram. Milan bought 600 g of sausages. Did he pay $7.74, $0.74, $17.74 or $18.90? Which answers are the best estimates for these problems? How do you know?

Ask the students to work in pairs to discuss their estimation strategies. Encourage the students to then calculate the answer. Calculators may be used.

Number card addition (1)

Materials

- a set of decimal cards labelled 0.9, 0.7, 0.4, 0.32, 0.45, 0.67

Choose two students to select a card.

What is the answer if you add these two decimals? (0.45 add 0.32, that is 70 hundredths add 7 hundredths . . . 77 hundredths.)

Discuss the students' strategies.

Variation

Repeat the activity using different decimal cards such as 0.5, 1.2, 2.5, 0.62, 1.32, 1.49.

Number card addition (2)

Materials

- a set of decimal cards labelled 0.132, 0.247, 0.295, 0.467, 0.475

Choose two students to select a card.

What is the answer if you add these two decimals? (That is 295 hundredths add 475 hundredths . . . 775 hundredths minus 5 hundredths, that is 0.770.)

Discuss the students' strategies.

Variation

Repeat the activity using different decimal cards such as 0.081, 0.801, 0.810, 0.018, 0.880 and 0.7, 0.708, 0.72, 0.725, 0.791.

Number card subtraction (1)

Materials

- a set of decimal cards labelled 0.9, 0.7, 0.4, 0.32, 0.45, 0.67

Choose two students to select a card.

What is the answer if you subtract the smaller number from the larger number? (0.4 subtract 0.32, that is 40 hundredths subtract 32 hundredths . . . 8 hundredths.)

Discuss students' strategies. Note that the preferred language for examples such as 0.32 is always '32 hundredths' rather than 'point three-two'.

Variation

Repeat the activity using different cards such as 0.5, 1.2, 2.5, 0.62, 1.32, 1.49.

Number card subtraction (2)

Materials

- a set of decimal cards labelled 0.132, 0.247, 0.295, 0.467, 0.475

Choose two students to select a card.

What is the answer if you subtract the smaller number from the larger number? (0.295 subtract 0.247 . . . You can count on, that is 3 hundredths to 0.250, then 45 hundredths, so 48 hundredths or 0.480.)

Discuss the students' strategies.

Variation

Repeat the activity using different cards such as 0.081, 0.801, 0.810, 0.018, 0.880 and 0.7, 0.708, 0.72, 0.725, 0.791.

Multiplying decimals

Materials

- a class set of calculators

Show these problems:

1.32km × 5 $2.65 × 4 0.89c × 6 0.28m × 8 7.45L × 3 0.675g × 3

How would you solve these problems?

Suggest students work in pairs and use rounding and estimating strategies. Encourage the students to then calculate the answer. Calculators may be used.

Now if you remove the decimal points from these problems and solve them again what do you notice? (The answers are the same but without the decimal point.)

Decimal problems can be solved by removing the decimal point, completing the calculation and then inserting the decimal point to the answer.

Dividing decimals

How much would it cost for one of each item?

Allow students to work in pairs and discuss strategies. When dividing decimals, it may be easier to remove the decimal point, carry out the division and then place the decimal point back at the end.

Multiply by 10

Materials

- a place value chart

Show these number sentences:

75 × 10 = ______ **750 × 10 = ______** **7500 × 10 = ______**

Can you tell me how to work out these problems? (It is easy to multiply by 10. You just add a zero.) Why do we add a zero when we multiply by 10?

Discuss the students' responses. Using the place value chart, explain that multiplying by 10 shifts each digit one place to the left.

× 10 ⟵ 1 place

Ten Thousands	Thousands	Hundreds	Tens	Ones
		7	5	0
	7	5	0	0

The number becomes larger by a power of 10 and a zero is inserted as a place holder.

Variation

Practise multiplying other 2- , 3- and 4-digit numbers by 10.

Multiply by 100 (2)

Materials

- a place value chart

Show these number sentences:

83 × 100 = ______ **431 × 100 = ______** **3256 × 100 = ______**

Can you tell me how to work out these problems? Why do we add 2 zeros when we multiply by 100?

Discuss the students' responses. With the aid of the place value chart, explain that multiplying by 100 shifts everything two places to the left.

× 100 ⟵ 2 places

Hundred Thousand	Ten Thousands	Thousands	Hundreds	Tens	Ones
		3	2	5	6
3	2	5	6	0	0

The number becomes larger by two powers of 10 and two zeros are inserted as place holders.

Variation

Practise multiplying numbers up to 6 digits by 100.

Multiply decimals by 100

Materials

- a class set of calculators

Enter 9 3 5 . 2 on the calculator.

What happens to this number if I multiply it by 100? (When you multiply by 100 it shifts everything two places to the left.)

Discuss the students' responses. Repeat the activity for these numbers.

57.424 × 100 = ______ **574.24 × 100 =** ______ **5742.4 × 100 =** ______

Variation

Practise multiplying multi-digit decimal numbers by 10 and 100 (powers of 10).

Decimal multiplication

Materials

- a place value chart

Show this number sentence: **6.85 × 1000 =** ______

Can you tell me how to work out this problem?

Discuss the students' responses. With the aid of the place value chart, explain that multiplying by 1000 shifts everything three places to the left.

× 1000 ⟵ 3 places

Thousands	Hundreds	Tens	Ones	Tenths	Hundreds
			6 •	8	5
6	8	5	0 •		

The number becomes larger by three powers of 10. Zero is inserted in the ones column as a place holder. Repeat the activity for other decimal numbers containing tenths, hundredths and thousandths. Note that calculators are a useful teaching tool in the development of decimal number patterns with decimals and may be used initially to increase motivation of students.

Divide by 10

Materials

- a place value chart

Show these number sentences:

500 ÷ 10 = ______ **5000 ÷ 10 =** ______ **50 000 ÷ 10 =** ______

Can you tell me how to work out these problems? (It is easy to divide by 10. You just cross out a zero.) Why do we cross out a zero when we divide by 10?

Discuss the students' responses. Using the place value chart, explain that dividing by 10 shifts everything one place to the right.

÷ 10 → 1 place

Ten Thousands	Thousands	Hundreds	Tens	Ones
5	0	0	0	0
	5	0	0	0

The number becomes smaller by a power of 10 and the zero is therefore deleted.

Variation

Practise dividing other 3- , 4- and 5-digit numbers by 10.

Divide by 100 (1)

Materials

- a place value chart

Show these number sentences:

2900 ÷ 100 = ______ **34 700 ÷ 100 =** ______ **631 800 ÷ 100 =** ______

Can you tell me how to work out these problems? Why do we cross out two zeros when we divide by 100?

Discuss the students' responses. With the aid of the place value chart, explain that dividing by 100 shifts everything two places to the right.

÷ 100 → 2 places

Hundred Thousand	Ten Thousands	Thousands	Hundreds	Tens	Ones
6	3	1	8	0	0
		6	3	1	8

The number becomes smaller by two powers of 10 and the two zeros are therefore deleted.

Variation

Practise dividing other 4-, 5- and 6-digit numbers by 100. Calculators may be used to investigate the resulting decimal numbers.

Divide by 100 (2)

Materials

- a class set of calculators

Enter [9] [7] [6] [3] [5] on the calculator.

What happens to this number if I divide it by 100? (When you divide by 100 it shifts everything two places to the right.)

Discuss the students' responses. Repeat the activity for these numbers: 524 324, 52 432.4, 5243.24.

Variation

Practise dividing multi-digit decimal numbers by 10 and 100 (powers of 10).

Decimal division

Materials

- a place value chart

Show this number sentence: **419 765 ÷ 1000 =** ______

Can you tell me how to work out this problem?

Discuss the students' responses. With the aid of the place value chart, explain that dividing by 1000 shifts everything three places to the right.

÷ 1000 ⟶ 3 places

Hrd Thousands	Ten Thousands	Thousands	Hundreds	Tens	Ones	Tenths	Hundredths	Thousands
4	1	9	7	6	5 •			
			4	1	9 •	7	6	5

The number becomes smaller by three powers of 10. Repeat the activity for other multi-digit whole numbers.

Decimal quantities

Show these statements: **0.2 of $200 10% of 150 kg 0.25 of 500 L 0.5 of $7500**

How would you calculate these decimal amounts? (You can convert to a fraction first: 10% is $\frac{1}{10}$ so to find $\frac{1}{10}$ of 150 kg, you divide 150 by 10.)

Discuss the students' strategies. Ask the students to devise some statements of their own. Practise calculating simple fractions, decimals and percentages of quantities often. Calculators may be used for more difficult examples.

Simple percentage

Show these statements:

10% of $500 20% of 200 km 25% of 1000 people 50% of $5000

How would you calculate these percentages? (You can convert to a fraction first: 20% is $\frac{1}{5}$ so to find $\frac{1}{5}$ of 200 km, you divide 200 by 5.)

Discuss the students' strategies.

Variation

Ask the students to investigate and identify the contexts in which fractions, decimals and percentages are used. Percentages and decimals are more commonly found than fractions.

Fraction notations

Show this information:

Fraction	Decimal	Percent
$\frac{1}{10}$		
	0.25	
		50%
$\frac{3}{4}$		
	0.2	
		70%
$\frac{6}{10}$		
$\frac{17}{100}$		
	0.05	
		100%

How can you work out the missing numbers? (0.25, 25 hundredths, that is zero point two five or 25%.)

Encourage students to explain their mathematical thinking when suggesting answers. Repeat the activity for fractions with denominators of 2, 3, 4, 5, 6, 8, 10 and 100.

Fractions, decimals and percentages

Show a set of cards with these fractions, decimals and percentages:

50%	0.99	$\frac{1}{10}$	0.05	$\frac{3}{4}$	33%	0.25	$\frac{2}{3}$	80%	0.4

How can we place these amounts in ascending order? (I imagined they were all fractions then put them in order. I thought of them all as decimals first.)

MONEY AND FINANCIAL MATHEMATICS

Year 5

Second-hand book sale

A Year 5 class have raised $80 for charity by selling second-hand picture books for $1.00 each, novels for $2.00 each, and magazines for 0.50c each. What combinations of these books could they have sold? (I think maybe 30 picture books, 20 magazines and 20 novels.)

If the class wanted to raise $20 more by selling picture books in 'buy two get one free' deal, how many books would they sell? *(You get 3 books for $2.00 and 10 × $2 equal $20.00, so 10 × 3 = 30 books.)*

Discuss the students' suggestions and look for a pattern.

Face paint stall

A face-painter at a market painted 25 full faces and 42 half faces. If a full-face cost $5.00 and half-face cost $3.00, how much money did the face-painter make? (25 × $5 is $125, $3.00 multiplied by 42 is $126, so $126 plus $125 = $251.)

Allow students time to use various strategies to calculate the costs.

If the face-painter had to pay $35 for the market stall and $65 on paints, how much profit did the face-painter make? (That is $251 less $35 then take away $65. That is $35 plus $65 – take it away from $251, that is $100, so $151.)

Extension

Ask the student to suggest strategies for how the face-painter could increase the profit.

GST

GST stands for Goods and Services Tax. It is a percentage of the cost added to the base cost of many goods and services to create the final price.

Ear buds $25.00 Ear pods $80.00 Headphones $15.00 Wireless headphones $45.00

*What would be the price of these items **after** a 10% GST is added?*

If you had $100 to spend and the store had a '10% off' sale, which items could you buy?

Discuss suggested purchases and calculations.

If your headphones were repaired at a cost of $18.00 before GST is added, what cost would you incur? (10% of $18 is $1.80, so $18.00 + $1.80 = $19.80)

Sandwich bar

What is the total cost of these sandwich orders, plus 10% GST – a salad sandwich for $5.50 and bottle of water for $3.00, a salad wrap for$6.00 and chocolate milkshake $5.50, a cheese sandwich for $4.50 and berry smoothie for $6.00, a salad roll for $5.50 and hot chocolate for $4.20.

Ask the students to create their own order using these items. They can work in pairs to calculate the total cost of the order, including the GST.

Year 6

10% off sale

A computer shop was having a '10% off' sale with discounts on laptops for $799.00, laptop cases for $60.00, office chairs for $65.00, office desks for $150.00, printers for $450.00 and printer inks for $70.00 each. What would the sale cost of each if these items be? What would you save if you purchased a laptop and printer? What would be your total saving if you purchased four printer inks? Which item will cost $15.00 less than usual in the sale? (To find 10% I moved the decimal point one place to the left. You find 10% then subtract that from the price.)

Discuss the students' strategies and suggestions.

50% off sale

A clothing store was having a '50% off' sale with discounts on t-shirts for $25.00, hoodies for $70.00, shorts for $30.00, track pants for $54.00, jeans for $110.00 and caps for $18.00. How do you calculate 50% off these prices?

Discuss the students' strategies and suggestions. Discuss how 50% is equivalent to 'half price'.

If you had $100.00 to spend, what would you purchase at the sale? How much would you save? (I would purchase a pair of jeans, a sweater and a cap costing $99, saving $99. I would buy 4 t-shirts, a pair of shorts and a hoodie costing $100 which saves me $100.00.)

25% off

If you go to the movies on a Tuesday, you can save 25% off the ticket prices: adults $24.00, children $16.00, students $18.00 and seniors $14.00. What strategies would you use to calculate 25% off these ticket prices? (I would calculate half or 50% then, half that amount again to get 25% then take that away from the price. I would divide the amount by 4 because 25% is the same as one-quarter, then subtract it from the price.)

Discuss the students' strategies and show them as number sentences, for example, **$24.00 – 50% = $12.00, $12.00 – 50% = $6.00, $24.00 – $6.00 = $18.00**.

Ask the students to calculate the cost of these ticket combinations at 25% off usual price: 2 adults, 1 adult and 2 children, 4 students and 2 seniors.

Variation

Ask the students to calculate the total cost for their own family to visit the movies on a regular day and on a Tuesday.

Percentage savings

At a sports clothing sale, runners that usually cost $220 now cost $165, compression tights that usually cost $120 now cost $60, crop tops that usually cost $45.00 now cost $40.50, fitness watches that usually cost $180 now cost $144, running shorts that usually cost $50 now cost $37.50, and t-shirts that usually cost $30.00 now cost $21.00. (The difference between $220 and $165 is $55, $55 is ¼ or 25% of $220. The saving on crop tops is $4.50, which is 10% of $45.00. $9.00 is saved on t-shirts, which is 10% × 3 so 30%. Compression tights are half price which is 50%.)

Students can work in pairs to calculate what percentage is saved for each item in the sale. Discuss the students' strategies.

PATTERNS AND ALGEBRA

Year 5

Four-digit puzzle

Show this number puzzle:

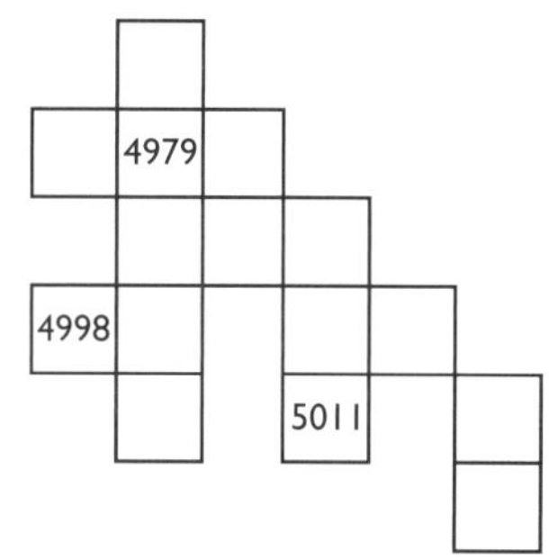

Can you find the missing numbers? How did you work out your answer? (There is a pattern. You count along by ones and down by tens.)

Variation

Repeat the activity for different-shaped puzzles and other 4-digit numbers.

Open puzzle

Show this open number puzzle:

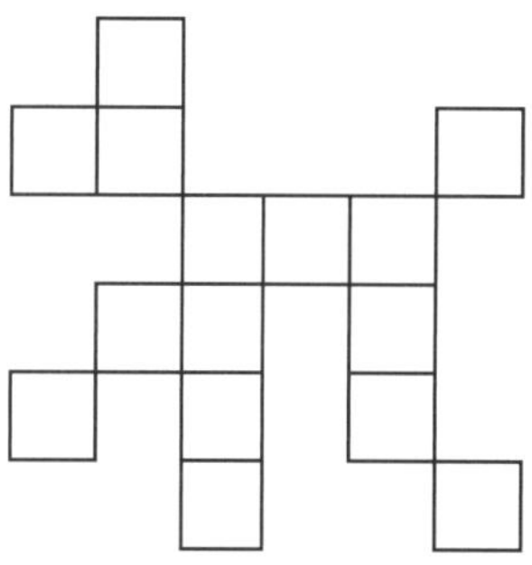

Can you suggest some missing numbers that will make the puzzle work? (You can write a number anywhere to start. You go along or down, forwards or backwards.)

Ask the students to design their own different-shaped pieces and insert 4- (or 5-) digit numbers.

Five-digit puzzle

Show this number puzzle:

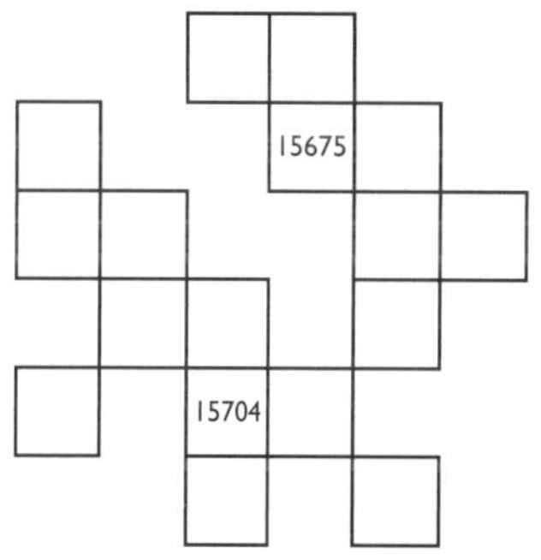

Can you find the missing numbers? How did you work out your answer? (You can add along by 1 and down by 10.)

Repeat the activity for different-shaped pieces and other 5-digit numbers.

Thousands jumps

Show an open number line segment from 75 000 to 85 000, with intervals of 1000:

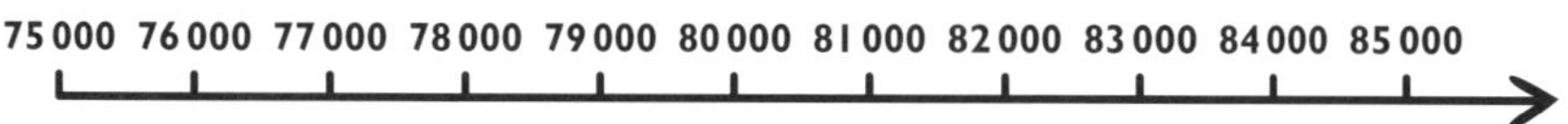

If I started at 75 000, how many jumps of 1000 would it take to get to 80 000? If I was at 81 000 and made 3 more jumps of 1000, where would I be? How many jumps of 1000 from 76 000 to 84 000?

Mark the jumps suggested by the students on the number line and discuss. Count by thousands from 75 000 to 85 000, forwards and backwards along the number line.

Square number patterns

Materials

- a multiplication grid

Show this multiplication grid:

×	1	2	3	4	5	6	7	8	9	10
1	1	2	3	4	5	6	7	8	9	10
2	2	4	6	8	10	12	14	16	18	20
3	3	6	9	12	15	18	21	24	27	30
4	4	8	12	16	20	24	28	32	36	40
5	5	10	15	20	25	30	35	40	45	50
6	6	12	18	24	30	36	42	48	54	60
7	7	14	21	28	35	42	49	56	63	70
8	8	16	24	32	40	48	56	64	72	80
9	9	18	27	36	45	54	63	72	81	90
10	10	20	30	40	50	60	70	80	90	100

What can you tell me about this pattern? (It is diagonal. They are square numbers.)

Discuss the square number pattern with the students.

Look at 16. The number to the top right and bottom left is the same, 15. These numbers are always one less than the square number.

Point out to the students that this knowledge allows us to recall 3 facts at once. For example, if 7 × 7 = 49, then 6 × 8 = 48 and 8 × 6 = 48, since 6 and 8 are on either side of 7 (6, 7, 8). List similar sets of 3 facts for other square numbers on the board.

Square relationships

Show this 10 × 10 grid:

What can you tell me about the square number pattern on this grid? (I can see a lot of square arrays on the grid. There are squares inside squares. The arrays are overlapping. You add an odd number to each square number to get the next square number.)

1									
	4								
		9							
			16						
				25					
					36				
						49			
							64		
								81	
									100

Discuss the students' responses. Develop the pattern of the square numbers.

1 + 3 = 4

1 +3 + 5 = 9

1 +3 + 5 + 7 =16

1 +3 + 5 + 7 + 9 = 25

1 +3 + 5 + 7 + 9 + 11 = 36

1 +3 + 5 + 7 + 9 + 11 + 13 = 49

1 +3 + 5 + 7 + 9 + 11 + 13 +15 = 64

1 +3 + 5 + 7 + 9 + 11 + 13 + 15 + 17 = 81

1 +3 + 5 + 7 + 9 + 11 + 13 + 15 + 17 +19 = 100

What is the rule?

Show this table:

A	B	C
6	7	42
8	9	72
7	8	
6	6	
9	9	
7	7	

Can you complete the missing numbers in the table? How did you get your answer? (If you multiply A and B you get C: 7 sevens are 49, add 7 equals 56, so 7 × 8 is 56.)

Discuss the students' strategies.

Extension

Ask the students to suggest how the rule could be written using symbols (A × B = C).

Add them up

Show this table:

A	B	C
13	15	28
26	14	40
84	36	
123	57	
344	460	
498	504	

Can you complete the missing numbers in the table? Which rule did you follow to get your answer? (If you add A and B you get C: take 2 off 504 and add it to 498, that is 502 add 500, that is 1002.)

Discuss the students' strategies.

Extension

Ask the students to suggest how the rule could be written using symbols (A + B = C).

Balancing multiplication

Show a balance scale with the multiplication number facts 9 × 4 on one side, and _______ × 2 on the other.

Which number will make the balance level? (You must get 36 . . . 2 eighteens are 36.)

Emphasise the idea of balance between both sides of an equation. The equals sign indicates 'is the same as', as opposed to a signal to perform a certain operation.

Variation

Challenge the students to devise equations of their own.

Balancing division

Show a balance scale with the division number facts 72 ÷ 8 on one side and _______ ÷ 9 on the other side.

Which number will make the balance level? (You must get an answer of 9 . . . 81 divided by 9 is 9.)

Emphasise the idea of balance between both sides of an equation.

Variation

Challenge the students to devise equations of their own.

Mixed balancing

Show a balance scale with the expression 10 + ______ on one side and 34 – 9 on the other side.

Which number will make the balance level? (You must get an answer of 23, so 10 plus 13 is 23.)

Discuss the students' strategies.

Variation

Repeat the activity for examples such as **6 × ______ = 66 – 12** **______ ÷ 4 = 45 + 5**

Open balancing

Show a balance scale with the expression ______ + 18 on one side and 4 × ______ on the other side.

Which numbers will make the balance level?

Discuss the students' strategies.

Variation

Repeat the activity for examples such as **72 ÷ ______ = ______ – 8** **4 × ______ = 54 ÷ ______**

Ask the students to devise some examples of their own and share them with their classmates.

Missing addition

Show these number sentences: **______ = 1.7 + 2.5** **2.4 + ______ = 5.2**

______ + 6.3 = 10.1

What do each of these number sentences say? How would you solve them? (2 and 4 tenths add something equals 5 and 2 tenths. You can add 3 and take off 2 tenths . . . 2 and 8 tenths.)

Discuss the students' strategies. Relate explanations to a number line segment. Often the equals sign is seen by students as merely an indication to perform an operation. Writing number sentences in the reverse form and with missing addends helps students to understand the idea of an equation. Practise examples such as the ones above often.

Missing subtraction

Show these number sentences. **______ = 6.8 – 3.6** **8.4 – ______ = 2.3**

______ – 10.5 = 23.6

What do each of these number sentences say? How would you solve them? (If I subtract 2.3 from 8.4, I will get the answer. It is 6.1.)

Discuss the students' strategies. Repeat the activity using similar examples. Ask the students to construct examples of their own.

Missing multiplication

Show these number sentences. **______ = 5 × 2.1** **8 + ______ = 8.8** **______ + 4 = 22**

What do each of these number sentences say? How would you solve them? (The answer is 1.1, that is 8 × 1 is 8 and 8 × 0.1 is 0.8, that is 8.8.)

Discuss the students' strategies. Repeat the activity using similar examples.

Missing division

Show these number sentences: ______ **= 25.5 ÷ 10** **7.7 ÷** ________ **= 1.1**

______ **÷ 4 = 4.1**

What do each of these number sentences say? How would you solve them? (7.7 divided by something equals 1.1. I will try 7.)

Discuss the students' strategies. Repeat the activity using similar examples. Calculators may be used to check answers.

Think of multiplication

Show these number sentences:

125 ÷ 5 = ______ **420 ÷ 6 =** ______ **350 ÷ 7 =** ______

What do each of these number sentences say? (It is saying 125 divided by 5.) Can you solve it using multiplication? (Yes, it is saying something multiplied by 5 equals 125, that is 25.)

Repeat the activity using similar examples.

Double and add

I am thinking of a number so that when I double it and add 10 the answer is 34. What is my number? (If you take away the 10, you get 24. That is double the original number. The number must be 12.)

Discuss the students' strategies. Show a number sentence according to the students' suggestions:

12 × 2 + 10 = 34 or 2 × 12 + 10 = 34

Variation

Repeat the activity using similar examples.

Halve and subtract

I am thinking of a number so that when I halve it and take away 20 the answer is 60. What is my number? (If you add 20 to 60 you get 80. That is the number halved. The original number must be 160.)

Discuss the students' strategies. Show a number sentence according to the students' suggestions:

160 ÷ 2 – 20 = 60 or $\frac{1}{2} \times 160 - 20 = 60$

Ask the students to devise similar word problems and express them as number sentences.

Year 6

Six-digit puzzle

Show this number puzzle:

Can you find the missing numbers?
How did you work out your answer?
(There is a pattern. You count along by ones and down by tens.)

134 975

134 994

134 007

Repeat the activity for different-shaped pieces and other 6-digit numbers.

Open puzzle

Show this open number puzzle:

Can you suggest some missing numbers that will make the puzzle work? How did you work out your answer? (You can write a number anywhere to start. You go along or down, forwards or backwards.)

Ask the students to design their own different-shaped pieces and insert multi-digit numbers.

Calendar patterns

Show this 4 × 4 grid. Tell students that the grid has been cut out from a calendar:

8	9	10	11
15	16	17	18
19	20	21	22
23	24	25	26

What patterns can you see in the grid? How would you describe your pattern?

Discuss the students' responses. *(If you take a 2 × 2 block, the differences vary by 2. For 3 × 3 blocks, the middle number is always the average of the 9 numbers. Each diagonal has a total of 71.)*

Calculators may be used during this investigation to assist motivation.

Window panes

Materials

- matchsticks

Show this arrangement of matchsticks:

Tell the students that the matches represent a row of windows.

How many matches are there in a window with 6 squares? How do you know? (6 squares will have 19 matches. There is a rule, you multiply the number of squares by 3 and then add 1.)

Discuss the students' responses. Label the first 6 squares in the pattern as 4, 7, 10, 13, 16 and 19.

How many matches are there in a window with 10 squares? 100 squares?

Discuss the students' responses.

Horseshoe patterns

Materials

- matchsticks

Show this arrangement of matchsticks:

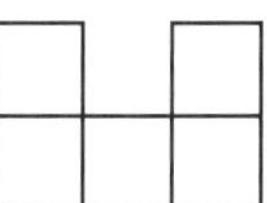

How many squares are there in each of these shapes?

Discuss the students' responses. Label the shapes as 5, 8 and 11.

How many squares are there in the next 3 shapes in the pattern? How did you work out your answers?

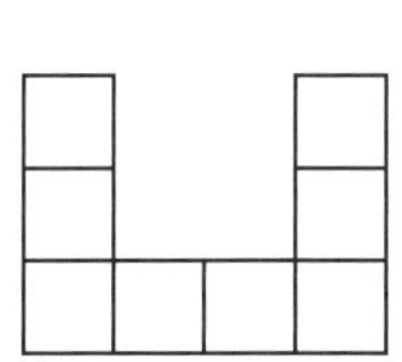

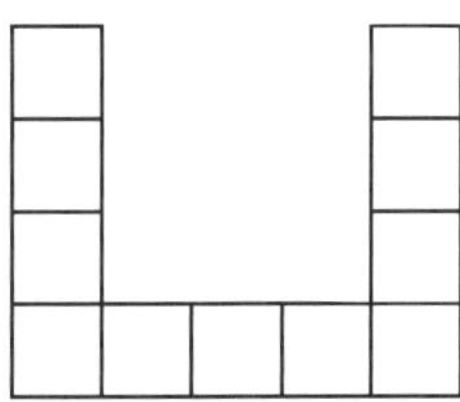

Discuss the students' responses. Develop the notation for each term in the pattern.

1 + 2 + 2   2 + 3 + 3   3 + 4 + 4   4 + 5 + 5   5 + 6 + 6   6 + 7 + 7

Letter L patterns

Materials

- matchsticks

Show these arrangements of matchsticks:

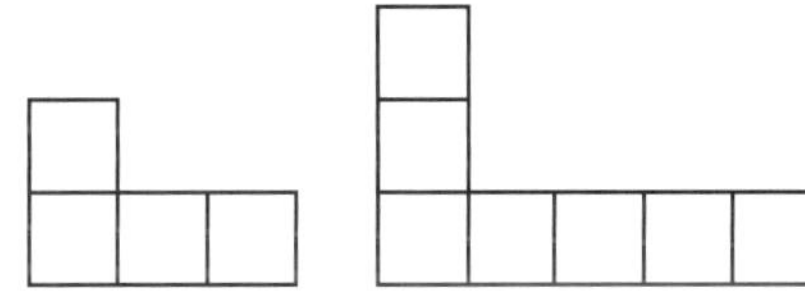

How many matchsticks are there in each of these shapes? (4 squares have 13. There is a rule. Each new shape has 9 extra.)

Discuss the students' responses. Label the shapes as 4, 13 and 22.

Pentagonal patterns

Materials

- counters

Show this pattern:

What are the next 2 shapes in the pattern? How did you work that out? (It is a pattern of triangles and squares. The next term is a 5 × 5 square with 4-3-2-1 on top.)

Label the first four numbers in the pattern as 1, 5, 12 and 22. Focus on the geometric structure of triangles on top of squares and groups of counters making the sides of a pentagon.

Tennis ball towers

Materials

- 20 tennis balls

Make the above pattern with tennis balls and show the students.

What are the next 3 arrangements in the pattern? How did you work that out? (The next term is 4 × 4, plus 3 × 3, plus 2 × 2 plus 1.)

Label the first five numbers in the pattern as 1, 5, 14, 30, 55 and 91. Develop the notation for each term in the pattern.

1 × 1   2 × 2 + 1   3 × 3 + 2 × 2 + 1   4 × 4 + 3 × 3 + 2 × 2 + 1

5 × 5 + 4 × 4 + 3 × 3 + 2 × 2 + 1   6 × 6 + 5 × 5 + 4 × 4 + 3 × 3 + 2 × 2 + 1.

Focus on the pattern of square numbers within the sequence.

Higher terms

Show this number pattern: **15 30 45 60 75**

What is the next term in the pattern? How did you work that out? (It is the pattern of 15s. You add 15 each time.) What is the tenth term in the pattern? (For the second term you multiply two 15s. For the 10th term it must be ten 15s.) What is the hundredth term in the pattern?

Discuss the students' mathematical thinking strategies.

Double plus one

Show this number pattern: **1 3 7 15 31**

What is the next term in the pattern? (It is double plus 1. Double 15 is 30 plus 1 is 31.)

Invite students to work in pairs to discuss their thinking. Show 63 as the sixth term.

What are the next 4 numbers in the pattern? (127, 255, 511, 1023)

Variation

Repeat the activity for other patterns such as **1 6 16 36 76**

Include examples with missing terms such as **1, 4, ___, 16, 25 and 1, 5, 13. ___. 61 125**

What is my rule?

Show this table:

Can you complete the missing numbers in the table? How did you get your answer? (If you divide A by B, you get C. That is 640 ÷ 8 = 80.)

Discuss the students' strategies.

How could you write this rule using symbols?

Discuss the students' responses. Develop the notation A ÷ B = C.

A	B	C
720	9	80
480	4	120
640	8	
560	7	
630	9	
540	6	

Mixed operations

Show these number sentences:

9 + _______ = 7 × 8 **_______ – 8 = 21 ÷ 7** **11 × 9 = _______ + 18**

127 – 46 = 9 × _______

What do each of these number sentences say? How would you solve them?

Discuss the students' strategies. Repeat the activity using similar examples.

Double and add

I am thinking of a number so that when I double it and add 46 the answer is 124. What is my number? (If you take away the 46 you get 78. That is double the original number so the number must be 39.)

Discuss the students' strategies. Show a number sentence according to the students' suggestions:

124 = _______ × 2 + 46 or 39 × 2 + 46 = 124

Variation

Repeat the activity using similar examples.

Quarter and subtract

I am thinking of a number so that when I take a quarter of it and subtract 15 the answer is 25. What is my number? (If you add 15 to 25 you get 40. That is $\frac{1}{4}$ of the number. The original number must be 4 × 40.)

Discuss the students' strategies. Show a number sentence according to the students' suggestions:

$$160 \div 4 - 15 = 25 \quad \text{or} \quad \frac{1}{4} \times 160 - 15 = 25$$

Ask the students to devise similar word problems and express them as number sentences.

Brackets

Show these number sentences: **$(9 \times 9) - (19 + 7) \times 2 =$ ______ ;**

$9 \times 8 \div (6 + 6) + 8 =$ **$(9 \times 4) \div (9 + 3) =$** **$\frac{1}{2}$ of $96 - (48 \div 8) + 3 =$**

Follow the rules of BODMAS (Brackets, Order, Division, Multiplication, Addition, Subtraction) to solve these number sentences. If you were to remove the brackets, would the answers remain the same?

Allow students time to solve the number sentences without brackets and compare. It is important to follow the BODMAS rule when solving number sentences with brackets – if the process is altered by removing brackets, answers can sometimes differ.